DS TRANSPORT

British Design

British Design

Hugh Aldersey-Williams

MoMA

Designed by Dondina Associati

The Museum of Modern Art
Editor
Rebecca Roberts

5 Continents Editions
Editorial Coordinator
Laura Maggioni

Layout
Annarita De Sanctis

Production Manager
Enzo Porcino

This book is typeset in Neue Helvetica.
The paper is R4 Matt Satin 170 grm².

Colour separation by
Eurofotolit, Milan

Printed and bound
in March 2010 by Conti Tipocolor,
Florence, Italy

5 Continents Editions
Piazza Caiazzo, 1
20124 Milan, Italy
www.fivecontinentseditions.com

ISBN: 978-88-7439-539-2

Distributed in the United States
and Canada by D.A.P./Distributed Art
Publishers, Inc., New York
www.artbook.com
Distributed outside the United States
and Canada, excluding France
and Italy, by Abrams UK, London

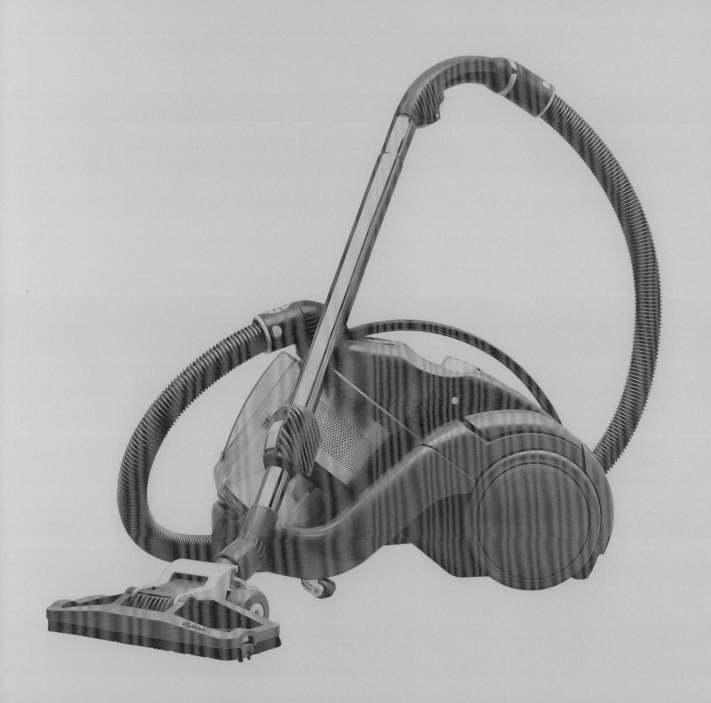

Paola Antonelli

A Strange Kind of Beauty

Despite the rather un-British boorishness of the slogan "Cool Britannia"—or perhaps because of it—one could not help but admire Great Britain's cultural renaissance of the 1990s. With the growing influence of London's annual 100% Design fair (founded in 1995), the waves made by the Royal Academy's 1997 traveling exhibition *Sensation*, the international success of such unusual design consultancies as Tomato and Mother, the inauguration of the Millennium Dome and Glasgow's Year of Architecture and Design in 1999, and the opening of Tate Modern in 2000, Great Britain deservedly began the new millennium at the center of the design world. A decade into the new century, British design is less bombastic, more socially engaged, and still uniquely farsighted.

In Britain, avant-garde quickly becomes mainstream, and the many attempts made by artists and designers to shock the middle classes have rapidly become aesthetic paradigms. As Hugh Aldersey-Williams writes in his essay in this volume, British culture is characterized by "moderation and compromise." High and low, Tory and Labour, old and new, European and colonial, tasteless and refined coexist and thrive in an unparalleled cultural synthesis. From Josiah Wedgwood in the eighteenth century and William Morris in the nineteenth to the Independent Group, the architects of Archigram, and the punk movement in the twentieth (not to mention Alexander McQueen), Great Britain has transformed the bipolar nature of its social and political structure into new, strange kinds of beauty.

This successful creative tension is supported by three traditions, which have given British design much of its special flavor. The first is economic rationality. Mass production was born in England in the eighteenth century, when Wedgwood first divided his ceramic ware production between two facilities—one factory for low-end, big-number series and another for more precious, one-off series. England is also the place in which mechanization first took command, in the second half of the nineteenth century, and where its limits were first discussed. The Arts and Crafts movement's return to traditional craftsmanship and its adherents' belief in "truth to materials" was an early instance of a critical approach to design that is in evidence today in the work coming out of the Royal College of Art's Design Interactions program, for example.

The second tradition is structural engineering, which, like politics, has long been considered a branch of aesthetics in Britain. This is demonstrated by the definition given in every issue of that country's Institution of Structural Engineers' journal: "Structural engineering is the science and art of designing and making, with economy and elegance, buildings, bridges, frameworks and other similar structures so that they can safely resist the forces to which they may be subjected."[1] This concept of engineering has made possible the innovative buildings sprinkled around the world by Norman Foster, Richard Rogers, Zaha Hadid, Foreign Office Architects, and many other architects based in England. They realized their projects in collaboration with such creative engineers as Cecil Balmond and the late Andrew Rice, both of Ove Arup and Partners, who are venerated as masters of architecture in their own right. The same structural daring, coupled with the ability to find the sublime in the most technological and mechanical aspects of human production, has had a powerful impact on industrial design. The work of several masters—Philip Vincent, James Dyson, Jonathan Ive—is a testament to this peculiar trait of British culture.

Finally, like British engineers, the British government has tended to acknowledge the structural forces behind creative phenomena, and since the middle of the nineteenth century—most notably the 1851 Great Exhibition in London and the consequent founding of the Victoria and Albert Museum—it has recognized the importance of design as a fundamental part of culture and of national interest in education and industry. The Design Council, founded in 1944, was one of the first government-based institutions in the world to promote design inside and outside its country. Despite the council's many ups and downs, it has contributed to the strength of British design and its image worldwide, funding and producing exhibitions and programs, publishing books and journals, producing statistics and other quantitative data highlighting the positive impact of good design, and encouraging design education starting in childhood, among other initiatives.

Because of its ability to both reflect and anticipate the pulse of culture throughout the industrial era, Great Britain has always represented a treasure trove for design curators at The Museum of Modern Art, who seek, as founding director Alfred H. Barr, Jr., intended, the "art of our time." Curators have followed British design and architecture with evident appreciation, as this volume amply demonstrates. The Museum's collection is rich in both historical and contemporary examples, among them objects currently at the forefront of design practice worldwide.

Indeed, some of the most inspiring and experimental designers and design firms in the world—Dunne & Raby, Hussein Chalayan, Paul Cocksedge, Thomas Heatherwick, Troika, and many others—live and work in London, and the Royal College of Art and Central Saint Martins College of Art and Design in that city continue to attract attention and draw renowned instructors. Manchester, the epicenter of the design world in the 1980s (the golden years of New Order), and Glasgow, whose 1999 festival reinvigorated a revolutionary design history, both continue to thrive. British tradition supports idiosyncrasy, appreciates cultural conflict, and embraces provocation when it is supported by real ideas and by technical mastery. Unless this changes, Great Britain's material culture will continue to provide the world with critique through design, prescient technological innovation, and its own strange kind of beauty.

1. Henry Petroski uses this example in *To Engineer Is Human* (New York: St. Martin's Press, 1985), p. 40. The editors of the journal *The Structural Engineer* include this definition in their "Mission Statement," www.istructe.org/thestructuralengineer/missionstatement.asp.

Hugh Aldersey-Williams

Finding Roles: Britain and Design

Histories of design in Britain conventionally begin with the Industrial Revolution, that rapid sequence of innovations in manufactures and organization of labor that occurred mainly in the English Midlands during the last half of the eighteenth century and the early part of the nineteenth. But to appreciate the aesthetic rules that govern British design it is helpful to go back further. In 1660, after a bitter civil war and a decade of Commonwealth government under Oliver Cromwell, the English monarchy was restored. Cromwell and his Roundheads had succeeded in strengthening parliament but had failed in their mission to bring about permanent republican government—they might have fared better had they shown the religious toleration they found wanting in their Catholic persecutors. The new king, Charles II, rejected the puritanism of the parliamentary regime but introduced some checks on royal excess in response to the public mood. This volatile period assuaged, seemingly for good, any national appetite for violent change and sowed the seeds of moderation and compromise that have characterized British life and taste ever since.

A few years after the Restoration, architect Christopher Wren was invited to devise a plan for the rebuilding of London after the twin catastrophes of the plague and the Great Fire. His comprehensive design called for the creation of nothing less than a new Rome in place of the tangle of narrow medieval lanes that had enabled both plague and fire to spread. Wren imagined great vistas, squares, and circuses. Portions of his grand scheme were realized—notably St. Paul's Cathedral and many fine new churches. But the plan that would have remade the nation's capital as an ideal city failed to win sufficient parliamentary support, probably because the wholesale razing and rebuilding it called for smacked too much of royal absolutism.

That the Industrial Revolution happened first in Britain is due to a unique convergence of factors: the innovations of Enlightenment scientists, the availability of markets and raw materials due to an extensive maritime trade, and the abundance of key native resources, notably the coal that would power the steam engines and feed the blast furnaces that would drive the looms and run the foundries of the new age. The key innovations of the period—the spinning machines of Richard Arkwright and James Hargreaves, the steam engine, and the steam locomotive—brought massive social upheaval as well as new opportunities. People moved off the land and into the cities, and an emergent middle class was able to afford artifacts, made by industrial processes, that were equivalent in function and appearance to the crafted items hitherto enjoyed only by the wealthiest.

The beginnings of industrial manufacturing necessarily initiated the activity of industrial design, although it would not be described in quite these words for another 150 years. Somebody had to think how to design objects in such a way that the new factories could mass-produce them. In *The Wealth of Nations,* of 1776, Scottish philosopher and economist Adam Smith marveled that an item as simple as a pin was now made in eighteen separate stages, from drawing the wire to forming the head to attaching the pins to paper for distribution.

The exemplar of early British industrial manufacture was the Wedgwood Pottery, in Staffordshire. Josiah Wedgwood was a Quaker with close ties to a group of innovators that included James Watt, the inventor of the steam engine. (Unable to swear an oath to the crown, Quakers of ability often went into commerce rather than law or government service.) After a conventional apprenticeship with one of the greatest masters of the time, Wedgwood set up his pottery in 1759. There he reshaped both the production and marketing aspects of the traditional craft. He drove up quality control, experimented tirelessly with glazes and other materials, and was attentive to good advice about style, often coming from aristocratic friends freshly returned from the Grand Tour of classical Europe. Wedgwood did not design or make the pieces himself, but he was close to both processes and managed to introduce mass-manufacturing methods, such as the use of transfers in place of hand painting, while retaining the essence of the craft. By offering customers a choice of decorative patterns that could be

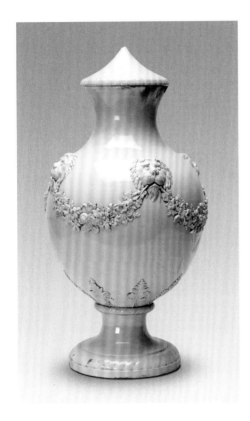

applied to a number of basic ceramic designs, he both increased the market appeal of his products and reduced his overhead.

Much Wedgwood china—Queen's Ware, for example—is Neoclassical, in tune with the general preference of the period, but some, such as the black basalt ware, bears comparison with modernist design of two centuries later. The radical new look was retrospectively justified by Etruscan precedents, a source of ideas that led in turn to further new products. No competing ideologies underlay these apparently contrary directions. The mercantile pragmatism of the Quaker mind was happy to pursue useful and ornamental styles alike, if both found a market.

The Georgian period matured into an extraordinary stylistic pluralism that ranged from the stripped classicism of military architecture to the icing-sugar plasterwork confections of domestic interiors. These may be seen with hindsight as representing two poles of the British character—the rationalist, puritanical, Enlightenment side, which retained a suspicion of ostentation, countered by the freer spirit of Romanticism.

One might expect the former to have prevailed when it came to designing the new Houses of Parliament in the 1840s, but Charles Barry and Augustus Pugin's winning proposal is a High Gothic extravaganza. It is perhaps emblematic of the dual nature of the British character that the national democratic legislature sits in a building that harks back to feudal times. The inventive classicism of John Soane, another competition entrant, would have offered a very different prospect. The oscillation between these two extremes was the key to the arts in the Victorian period, many of the finest works of which—such as

George Baxter (British, 1804–1867). *The Exterior of Crystal Palace, Sydenham*. Nineteenth century. Color lithograph, dimensions unknown. Maidstone Museum and Art Gallery, Kent, England / Bridgeman Art Gallery

Joseph Paxton's Crystal Palace built for the Great Exhibition of 1851—succeeded by bringing about a synthesis of these contrary impulses. The Great Exhibition attracted six million visitors and turned a healthy profit, which was used to endow the schools of design and science and the museums of South Kensington, near where the Crystal Palace stood in Hyde Park.

The Arts and Crafts movement, its label derived from the title of an 1888 exhibition, was probably the first major stylistic movement to overtake what were then called "the useful arts" in Britain. It affected architecture, furniture, textiles, and graphic design. Inspired by an idea of Britain's lost rural arcadia, designers such as William Lethaby, C. R. Ashbee, and C. F. A. Voysey sought to keep a place in the growing world of mass production for the skilled blacksmith and carpenter and other craftspeople. As a movement motivated by social idealism more than formal considerations, it should have had no distinguishing style, but Arts and Crafts works are readily identifiable by their studied asymmetries of design and their juxtaposition of contrasting materials such as brick and tile or forged iron and carved oak. As a counterblast

against the Industrial Revolution, the movement was both politically ineffectual and morally conceited. However, the style struck a chord with the public and seemed to answer a rather unconsidered hankering for the preindustrial past.

The leading Arts and Crafts design theorist was William Morris—author of the 1890 utopian novel *News from Nowhere* and friend of the Pre-Raphaelite circle of artists. Morris put his socialist ideas into practice in printing, publishing, and manufacturing stained glass, wallpapers, and textiles. His detailed, nature-inspired patterns are no mere surface pleasures: "They are depths," William Lethaby said.[1] But in practice, Morris's designs were hardly less exploitative of labor than the alternatives, and most could only be afforded by the very rich. If Morris has since become something of a hero for the anticapitalist movement, its adherents need to look again at this independently wealthy son of a stockbroker. Conversely, his position as the whipping boy of machine-age technocrats is also undeserved—he was not the effective brake on industrial progress they suppose.

What to do about traditional crafts as they succumbed in turn to mechanization became an important problem for designers. It was clear that long-established processes and materials—wood carving in oak, for example—were no longer compatible with the requirements of the market, yet there remained a respect and a fondness for objects that appeared to show "honest workmanship." Pieces like Edward Welby Pugin's chair of 1870 (he was the son of the architect) are a response to this impasse. It demonstrates honesty through a visible structure designed along Gothic principles and even anticipates some of the stylistic conceits of the high-tech architecture of the 1970s and 1980s, in features such as the "go faster" holes drilled through its heavy timbers in an effort to reduce weight.

There are moments in this history when a door briefly opens onto a quite different design landscape but the threshold is never crossed. Christopher Dresser was one of several creators whose work suggests that British design might have taken a much more radical, modernist turning at this time; his metalwork in particular is minimalist and angular, anticipating the Bauhaus by forty years. Although Dresser had many inspirations and moral prejudices in common with his contemporaries—a fondness for Japanese art, a preference for "honest" design—his early training in botany (he acquired a number of professorships), which required him to produce illustrations of scientific accuracy, showed him how to take the ethos to a new extreme of rigor. The trace of handicraft was not important for Dresser as it was for so many of his contemporaries—what mattered more was to achieve a level of platonic perfection whatever the material and end use of the object. Although, like Morris, he proselytized for his particular design philosophy, his call was not widely heeded.

It is appropriate here to pause to consider the international context in which British designers operated at this time. Despite, or perhaps because of, the presence of the British Empire, which was at its greatest extent and strength near the turn of the century, designers in Britain by and large felt little compunction to stay abreast of trends on the Continent. Their influences came either from a (partly imagined) glorious national past or, in vicarious fashion, from far-flung corners of the globe over which the British crown exercised dominion. This did give them plenty of scope: styles ranged extravagantly from mock-Tudor beams and Gothic arches in suburban houses and churches to Mogul-style lamps and jewelry. The Arts and Crafts movement had been convened in an attempt to superimpose an autochthonous new style over all this historicist chaos, but of course it simply added one more look to the plurality.

Important movements such as Secessionism in Vienna and, closer to home, the Art Nouveau that was most strongly expressed in France and Belgium had very little impact on the mainstream of British design and caught the eye of a very few. One of those few was Aubrey Beardsley, whose erotic Orientalist illustrations were perfectly suited to Oscar Wilde's play *Salome*, which opened in Paris in 1896, but hardly to the mainstream of British taste.

Another was Glasgow-born Charles Rennie Mackintosh, who at a very early age demonstrated prodigious ability as an architect of great originality. The distinctive style that he developed—an improbable fusion of Art Nouveau, Japanese art, and Scottish Baronial—was applied with extraordinary success to buildings, furniture, textiles, and cutlery. His masterpiece was the Glasgow School of Art, completed in 1909. Mackintosh's talent was widely recognized on the Continent, and he bequeathed Glasgow a visual trademark that it has exploited to great advantage ever since.

The example of Mackintosh raises the issue of regionalism in British design. Though not as much as France, Britain is a centralized country, and if most of the cultural energy is not actually concentrated in London, there is certainly the assumption that it is. Industrial design emanated originally from the Midlands, where manufacturing was based. Other cities, notably Manchester, Sheffield, and Glasgow, have made a distinctive impression from time to time, but the long-term trend, accelerated in the postindustrial period, has been for what are now termed the "creative industries" to gravitate to the capital.

In the years from the end of Queen Victoria's long reign to the outbreak of World War I, Britain consolidated the colonial and technological gains of the nineteenth century and basked in unprecedented wealth and influence. Yet it was also a time of great inequality. The socialist future envisioned by Morris and others had certainly not been realized, and many people worked long hours for scant reward while their masters lived in luxury.

This upstairs-downstairs world could not last. Influential surveys revealed the extent of poverty in Britain. Membership of labor unions surged. Women fought for the vote. All this turmoil demanded a graphic language, and often the solution was to adapt the lettering and imagery of ubiquitous Victorian consumer advertising to political ends. Art Nouveau graphics, for example, were judged helpful "to undermine the 'ugly suffragette' theory and keep the tone soft yet persuasive," Liz McQuiston has written.[2]

Charles Rennie Mackintosh (British, 1868–1928).
Glasgow School of Art. 1909. View of the library

Opposite: Aubrey Beardsley (British, 1872–1898).
The Climax. One from a portfolio of illustrations for the
play *Salome*, by Oscar Wilde. 1894. Medium unknown,
8⅞ x 6⅜" (22.5 x 16.2 cm). The Museum of Modern
Art Library, New York

The slaughter of the First World War—more than thirty million casualties from two dozen combatant nations—showed the need for a new moral order. In various countries drastic political renewal was presaged by radical movements in design— Futurism in Italy, Constructivism in Russia, and the founding of the Bauhaus in Weimar, the birthplace of the new republic of Germany. The reaction in Britain was characteristically muted and ambivalent. Some Britons believed that life could go on as before; others realized that although little had happened to affect their material existence, everything had changed underneath.

Implicit acknowledgement of superior German industrial organization came with the advent of the Design and Industries Association (DIA) in 1915, modeled after the Deutscher Werkbund, which had been founded in 1907, but on the whole the DIA produced neither common cause nor constructive dialogue between the two factions it purported to represent. In general, industry was slow to see the advantages of mechanized production, and many designers continued to create objects reliant on handwork rather than rising to the new creative challenge.

After the war, the London Underground embarked upon a major program of renovation of existing subway stations and construction of new ones on lines extending into the suburbs. A relatively untried and apparently conventional Arts and Crafts–schooled architect, Charles Holden, was chosen for the work. But as he was given his head, a distinctive new style emerged, of light-flooded, cathedral-like spaces dedicated entirely to public use. These generous works by a single designer had the useful side effect of establishing a progressive house style for the Underground. Thus did a kind of modernism make its first

Edward Johnston (British, 1872–1944). London Underground Roundel. 1913

Opposite: Sir Giles Gilbert Scott (British, 1880–1960). Royal Fine Art Commission for Scotland (UK, established 1927). K2 Telephone Box, prototype model. 1925–30. Various materials, 35⅛ x 12³⁄₁₆ x 12³⁄₁₆″ (89 x 31 x 31 cm). Museum of London

appearance in the lives of a great many British people living in the expanding "Metroland" around the capitol.

The integration of London's subway system demanded a unified design that would embrace signage and advertising as well as architecture. The key was the now familiar circle-and-bar symbol and a unique typeface, both developed by Edward Johnston, an approach that was successfully emulated by some of Britain's main railway companies. The motoring industry, in the form of companies such as the gasoline distributor Shell, competed with its own graphic vision, which veered between the pastoral and the futuristic.

The graphic design that accompanied the expansion of the Underground illustrates the ambiguous status of design between the wars but demonstrates that, with the right support, excellent work could be done. Commercial artists designed posters advertising the virtues of the country-side opened up by the new outlying stations on the network, presenting a rural idyll to the urban masses. Meanwhile, the young draftsman Harry Beck took up—as a spare-time diversion, not a company commission—the daunting task of designing a route map that would present the Underground as a coherent system rather than the patchwork of former private companies'

fiefdoms that it was. Beck's extraordinarily felicitous solution, based on the conventions of electrical circuit diagrams, was an immediate success when tested on the public in 1933. Its schematic representation has been criticized for making outlying places seem closer to the center of the city than they are,[3] but this, to suburbanites, was exactly the effect of the subway itself.

As is the case so often in these situations, none of this extraordinarily enlightened patronage would have happened but for the presence of one man: Frank Pick. Pick was trained as a lawyer and rose as a manager through London Transport, but he had a passion for design. Unlike most managers who doodle their own designs, however, he knew when to hand a job over to a professional, and the professionals he chose and the freedom he granted them remain an example for design managers today. Under his guidance London Transport became what the architecture critic Nikolaus Pevsner called "the most efficacious centre of visual education in England."[4] Pick and Holden had met at the inaugural meeting of the DIA—if only that organization had been able to replicate their synergy throughout British industry.

Despite these advances in the public realm, people's homes for the most part continued to be crammed with heavy, drab furnishings essentially little different from those found in Victorian homes a generation before. For ambitious young designers these were dark days.

Cultural prospects were enlivened by the influx of designers emigrating from Nazi Germany and Stalin's Soviet Union. Not all the designers who came wished to make Britain their home, and many moved swiftly on to America. But Walter Gropius, the founder of the Bauhaus, and Marcel Breuer were among those who stayed long enough in the mid-1930s to make a real impact. At a stroke, Britain's insular self-satisfaction was shattered and the country was forcibly exposed to that most dangerous of things—Continental ideas. The clearest signal of this was perhaps the De La Warr Pavilion at Bexhill-on-Sea, East Sussex, England, designed by Serge Chermayeff and Erich Mendelsohn, which presented the most radical imaginable transformation of the cozy world of English seaside architecture with its clean white lines and simple geometries.

One of the biggest beneficiaries of this influx was the Isokon company of Jack Pritchard, which set about manufacturing Breuer's bent plywood furniture. The company was also responsible for the Lawn Road Flats in the London suburb of Hampstead, a white stucco apartment block built as a manifesto for modernist living. The class of its residents—sculptors

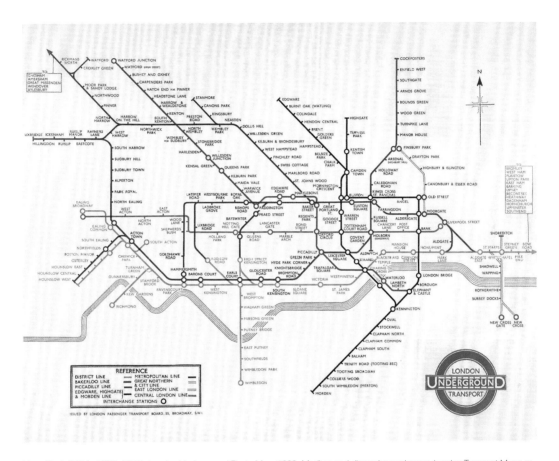

Harry Beck (British, 1902–1974). London Underground Route Map. 1933. Medium and dimensions unknown. London Transport Museum

Henry Moore and Barbara Hepworth and author Agatha Christie, among them—implies that such living was not yet for everybody.

Isokon was not an isolated case. Gerald Summers established a company called Makers of Simple Furniture and designed his remarkable Lounge Chair (1934), made from a single rectangular piece of bent plywood. Alvar Aalto was the obvious inspiration, but the design was more the solution to a self-set problem than a piece of furniture likely to appeal to the market. More commercially successful was Gordon Russell, who used the same materials and methods

Marcel Breuer (American,
born Hungary, 1902–1981).
Side Chair. 1936–37.
Walnut and birch plywood,
29¼ x 15¾ x 15½"
(74.3 x 40 x 39.4 cm).
Manufacturer: Isokon
Furniture Company,
London. The Museum
of Modern Art, New York.
Gift of Eliot Noyes, 1946

to produce furniture that was modern but not conspicuously avant-garde. Russell's eminent practicality saw him later put in charge of the Utility Furniture Advisory Committee during World War II, a position he exploited, Jeremy Myerson has written, to force the British to break with "the pre-1939, pseudo-Victorian tradition of reproduction furniture."[5]

The main architect of Isokon's Lawn Road Flats was Wells Coates, perhaps the only Briton (though born in Canada) to successfully emulate the example of men such as Walter Dorwin Teague and Henry Dreyfuss in American industrial design, most notably in his Bakelite casings for radios made by Ekco. Almost alone in British product design, his work was alive to the international trends of Art Deco and the American machine age. Contemporary with Coates's high-styling approach, George Carwardine's spring-counterbalanced Anglepoise Lamp of 1932 inspired many imitations and remains a popular archetype, one of the most successful examples of engineering-inspired design.

World War II was kind to design in Britain. Although people went without many consumer goods and even basic foodstuffs were rationed, designers found themselves in demand to help the war effort at various levels, from propaganda art to military engineering design. The beautiful Spitfire and the more effective Hurricane fighter aircraft and the Rolls Royce engines that powered them won an admiration that cannot be solely explained by their role in beating off the German aerial invasion of 1940. In parts of rural England it is a feature of the summer that one or two of these aircraft pass overhead on their way to or from some display, and the distinctive purr of that engine can produce a lump in the throat even in those too young to remember the war. Less lovable but quite as much of a breakthrough was the jet engine, for which airman Frank Whittle laid the groundwork as early as 1930. Crucially, perhaps, he was no engine builder but rather that vital stimulant of design, the dissatisfied user with the imagination to see that an improvement can be made.

This was the first war Britain had been obliged to fight on the home front, and so it is not surprising to find that government propaganda played a substantial role in encouraging the public to do its bit. What is curious is that the work produced—often by designers from the Continent, such as the German-born Frederick H. K. Henrion, who came to London from Paris in 1936—was of so much better quality than was needed to do the job. Equally curious is that many of the designs were only printed somewhat after the crisis they purported to address, in celebration or commemoration. Thus, Henrion's iconic image of a swastika pulled apart by the firm grip of the four Allied powers of Britain, France, the United States, and the Soviet Union was intended to be posted *after* the D-Day liberation.

In hindsight, it is easy to see that British voters would reject the values of the Conservative Party after the war, even though it had been that party's leader, Winston Churchill, who led the country to victory. In the 1945 general election they voted Labour by a landslide, hoping for an end to austerity (which they didn't get) and greater social justice (which they did). One of the new government's first actions was to set up the National Health Service to provide free universal health care paid for by taxation.

In 1944 the government had set up the organization that soon became known as the Design Council. Its task was to exhort British industry to make better use of design and of graduates from the country's flourishing design schools. Inspired by similar initiatives in Scandinavian countries, the Design Council published case studies in the use of design, funded companies too timorous themselves to pay for the services of design consultancies, and administered an awards scheme for good consumer design until well into the 1980s. (Surprisingly, the Design

Frederick H. K. Henrion (British, 1914–1990). *Untitled (Four Hands)*. 1941–45. Lithograph poster, 26¹⁄₁₂ x 19¹⁄₂″ (66.3 x 50.6 cm). Imperial War Museum, London

Abram Games (British, 1914–1996). Cover design for *The South Bank Exhibition, Festival of Britain: A Guide to the Story It Tells*, by Ian Cox (London: His Majesty's Stationery Office, 1951). Commercially printed book, approx. 11¹¹⁄₁₆ x 8¼″ (29.7 x 21 cm). Museum of London

Council still attracts government funding today, when the global economy exposes British companies to the true nature of the competition they face and gives designers the mobility to market their skills internationally.) It is hard to escape the impression, however, that those companies that made the best use of design in the early postwar period could work it out for themselves, while the majority of manufacturers, who didn't much care for good design, were unlikely to respond to the evangelical blandishments of a government agency.

The greatest cultural achievement of the first postwar government was undoubtedly the Festival of Britain, staged in 1951. Serendipitously falling on the centenary of the Great

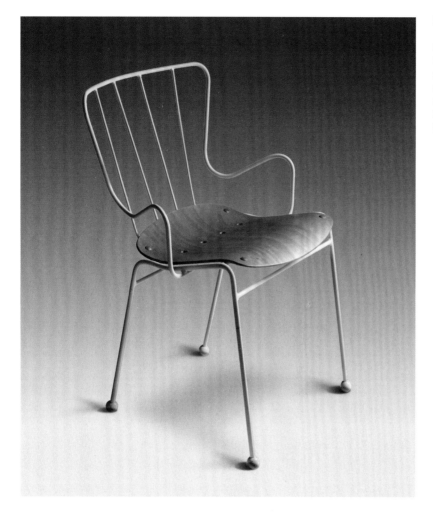

Ernest Race (British,
1913–1964). Antelope
Chair. 1950. Enameled,
bent steel rod and
molded plywood,
31⅞ x 20¹⁄₁₆ x 15¾"
(81 x 51 x 40 cm).
Manufacturer: Ernest
Race Ltd., England

Exhibition, the festival provided a tonic to the nation when it was most needed. The main London site was developed with extravagant structures, some of them demountable for potential use elsewhere and some, like the Skylon—an aluminum tensegrity beacon inspired by Buckminster Fuller—with no other purpose than to demonstrate their own delightful possibility. The focus for once was not solely on London: a traveling exhibition went to other major cities, and a festival ship called at national ports. Ten million people visited and came away with enduring memories of an event that showed true creative spirit—in marked contrast to the Millennium Dome fifty years later, which has drawn similar visitor numbers but left no positive cultural imprint.

The design on display at the Festival of Britain was a mostly successful marriage of Scandinavian modernism and a domesticated version of the lightweight, engineering-inspired structures that had become familiar during the war. The overall effect was to largely banish the national stereotype of "workmanlike" and "solid" design and to suggest that modern living was an option open to all, not just the eccentric upper classes or an urban aesthetic elite.[6] (Further ripples of this effect would be felt with the advent of two influential houseware retailers, Terence Conran's Habitat in the later 1950s and the Swedish Ikea in the 1980s.) For those who found that this was not their cup of tea and who could not bear, in the words of a future Ikea advertising campaign, to "chuck out the chintz," there was celebration of a more traditional kind with the coronation of Queen Elizabeth II in 1953—although thousands of households made the concession to modernity of buying or renting their first television in order to share in the spectacle.

The "new Elizabethan age," as this period was coyly labeled, saw modernism gain greater acceptance by the British people in a version shorn of ideological do-goodism and given an almost whimsical, light touch. Ernest Race's Antelope Chair, a curvaceous skeleton of enameled metal rods with a colorful molded plywood seat, is representative. The same slightly humorous, futuristic look is seen in many popular tableware and textile patterns of the time, such as Ridgway Potteries' Homemaker china and the work of the Festival Pattern Group, the latter derived, somewhat improbably, from the X-ray patterns recorded by crystallographers in groundbreaking research into the structure of biological molecules, such as the newly described DNA and insulin.

Another source of stylistic innovation lay in the work of craftspeople such as Lucie Rie and Hans Coper. Like Breuer and Gropius they had fled Nazi Europe, and like them they immediately challenged Britain's insular preconceptions, showing that craft was an activity involving brain as well as hand. Their ceramics moved craft away from design and closer to art—in their work functionality often takes second place to form. At the same time, cutlers Robert Welch and David Mellor were making new strides in integrating craft and mass manufacture in silver and steel, as David Queensberry and Martin Hunt did soon after in ceramic design. The old

Martyn Rowlands (British, born 1923). Sample of "Haemoglobin" Plastic Laminate. c. 1951. Laminated plastic, 18¹¹/₁₆ x 10⅜"
(47.5 x 26.3 cm). Manufacturer: Warerite Ltd., England. Victoria and Albert Museum, London

David Mellor (British, born 1930). Pride Cutlery. 1953. Silver plate, dessert spoon: 7¼ x 1⁷/₁₂" (18.5 x 4 cm), table fork: 8½ x ¹⁵/₁₆"
(20.5 x 2.4 cm), table knife: 8⁷/₁₆ x ¹¹/₁₆" (21.5 x 1.8 cm). Manufacturer: David Mellor Design Ltd., Sheffield

Arts and Crafts notion of craft as the handmaking of useful artifacts seemed at last to have been abandoned.

Although people could finally begin to forget the war, other shadows loomed in the 1950s. In the words of Dean Acheson, United States secretary of state from 1949 to 1953, Britain had "lost an empire but not yet found a role." This truth was demonstrated by the Suez Crisis of 1956, in which Britain and France, abetted by Israel, failed in a military attempt to prevent Egypt from nationalizing the Suez Canal, which they relied upon for oil and other imports.

Cover of sales brochure for the British Motor Corporation's Morris Mini-Minor car, 1959

The Mini, the tiny saloon car that was destined to become the automotive icon of the 1960s, was launched in 1959, partly in response to the increased cost of oil following the Suez debacle. Its design, by Alec Issigonis—a Greek-German refugee from Turkey—should be regarded as lying squarely within a British tradition of quirky engineering inventiveness, not as an attempt to design for mass-market fashion. Indeed, although Minis sold at a rate of one hundred thousand a year, it is said that the production line never made a profit.

It is always invidious to point to one automobile as the distillation of a national psyche, and in truth neither the Mini nor other oft-cited British cars of the period—the ostentatious Rolls Royce Silver Shadow (1965), the louche Jaguar E-Type roadster (1961), or the Morris Traveller (1952), with its half-timbered rear end—exactly fits the bill. A better symbol of Britishness is the national system of road signage created by Jock Kinneir and Margaret Calvert between 1957 and 1967, whose pragmatic, softened modernism has the thoroughly worked-out system but little of the visible dogma of Continental equivalents. The font they adopted was a version of Akzidenz Grotesk, modified with small cursive flourishes to improve legibility; the execution was largely but not entirely in lowercase and the background colors were chosen to harmonize with the landscape.

Typography is one design specialty in which the British willingness to strike a compromise has tended to prove a strength rather than a weakness. Kinneir and Calvert's font fits well with designs such as those by Eric Gill from the 1920s (Gill Sans and the Arts and Crafts–influenced Perpetua) and the continuing work of Matthew Carter, who has collaborated on versions of Helvetica and developed computer fonts such as Verdana.

The economic boom of the late 1950s, coming after so many years of austerity, found its cultural echo in the 1960s with an explosion of creative energy in all fields. It had begun in 1956 with the influential exhibition *This Is Tomorrow* at Whitechapel Art Gallery, London, in which Richard Hamilton and other British Pop artists paraded a guiltless celebration of American-style consumerism. Inspired by American rock and roll and blues, The Beatles and the Rolling Stones began performing in the early 1960s and soon became international sensations.

Design was at the center of the maelstrom—many of the new pop bands had their beginnings in art schools and, thanks to television, the visual element was now more important than ever before in this field. The result was the seamless fusion of fashion, music, and design captured in Michelangelo Antonioni's 1966 film *Blow-Up*. Improbably enough, frumpy Britain became an international center for accessible fashions, with designers such as Mary Quant

and retailers such as Biba, while Carnaby Street and the King's Road became places of tourist pilgrimage.

Advertising boomed in this new consumer age. A new generation of professional designers, not content to enslave themselves to a single manufacturer, began to form consultancies along the lines of advertising agencies. The trailblazer was Fletcher/Forbes/Gill, formed in 1962 by graphic designers Alan Fletcher, Colin Forbes, and Bob Gill, which grew into the multidisciplinary firm Pentagram ten years later with the addition of industrial designer Kenneth Grange and architect Theo Crosby. The partnership arrangement played to everybody's strengths, allowing the consultancy to design products and graphic materials for multiple clients while retaining complete creative independence. Work ranged from the Art Nouveau–inspired logo for Biba (by John McConnell, who joined in 1974) to Grange's Instamatic cameras for Kodak.

The business model was unlike that established in the United States a generation before by the first "industrial designers" called by that name. American design firms prospered from the legendary charisma and salesmanship of the creative individuals at their heads, whereas Pentagram and the other big British design consultancies that followed, such as Wolff Olins (founded 1965) and Moggridge Associates (1969), the forerunner of IDEO of London and California, relied on an aura of professional competence that tempered creativity with business acumen to win over skeptical clients. The situation differed also from that on the Continent, where designers more often worked individually or in small studios. The British concept of the full-service consultancy, whose designers could approach each client's problems afresh, was easily comprehended by business and provided designers with an environment of constant stimulus. Its success had a number of effects: the growth of a "design community," an even greater emphasis on London as the center of design in Britain, and the decline of in-house corporate design.

In the best of cases, manufacturing clients developed long-term relationships with a single consultancy. This happened with conspicuous success in the case of the kitchen appliance manufacturer Kenwood and Kenneth Grange. Taking his lead from the work of Dieter Rams for Braun, in Germany, Grange has worked closely with Kenwood since 1960, guiding the evolution of a mutedly progressive style across a wide range of products.

Elsewhere, innovative design continued to spring from the engineering tradition. Alex Moulton, Issigonis's onetime colleague on the Mini, devised a series of innovative small-wheeled bicycles geared to urban life, beginning in 1962. Owen Maclaren did the same for the baby buggy in 1965, bringing out the first folding stroller. What they often lacked in elegance, these designs made up for in functional ingenuity, a trade-off that British consumers are often more than happy to accept. Better this than the opposite, perhaps: Clive Sinclair rose to fame in the 1970s with digital gadgets whose assured minimalist styling, far ahead of its time, stands comparison with the products of Apple or Sony today—but his scientific calculators did not always get the right answer.

In 1973 another oil crisis brought a fresh reminder to Britons that their country was at the mercy of external forces. Glamorous projects such as the Anglo-French *Concorde* supersonic airliner and a tilting high-speed passenger train that had seemed plausibly ambitious at the close of the optimistic 1960s now looked vainglorious. The truth was that the country had enough trouble manufacturing a decent car.

The 1970s were characterized by industrial unrest. Pushed to the edge, many businesses simply failed to recover, and Britain's uncompetitive manufacturing sector began an agonized and inexorable decline. The chronic inability of the bulk of British industry to see that design offered an answer to their woes was undoubtedly a factor, though in truth the greater part of the problem lay with ossified industrial relations rooted in the class system. British consumers turned, reluctantly at first, but then with growing enthusiasm, to imports from the Continent and Japan.

The general malaise found expression in album cover art, one of the few areas of design to flourish, where the escapist fantasies of Hipgnosis and Roger Dean for bands like Pink Floyd and Yes gave way to the urgent typographic cries of Malcolm Garrett, Peter Saville, and Neville Brody, which spoke more directly of industrial decline and unrest. This graphic specialism continued to thrive until the demise of the twelve-inch record with the introduction of compact discs in the 1990s.

Perhaps a more characteristically British contribution to design at this time was the style of architecture known as "high tech"—seemingly inspired by the traditional British metal con-struction toy Meccano, in contrast to the postmodernists' blocky Lego. Its champions were Richard Rogers and Norman Foster. Though owing much to American precursors such as the Eames House of 1949, high tech in the hands of these young architects achieved a rare

Peter Saville (British, born 1955). Cover design (front and back) for record album *Power, Corruption & Lies*, by New Order. 1983. Commercially printed cardboard record sleeve, 12¼ x 12¼" (31.1 x 31.1 cm)

marriage of technology and craftsmanship that seemed uniquely British. This was in part by force of circumstances: the American aesthetic was built on standard parts ordered from catalogues, but such catalogues did not exist in Britain. It is notable that many high-tech architects' commissions came from abroad at first, following the soaring success of the 1977 Centre Pompidou in Paris (by Rogers, with Renzo Piano). But with Rogers's Lloyd's Building (1978–86) in London, high tech was soon embraced for corporate as well as cultural statements. The aesthetic of painted steel beams, aluminum panels, and glass with both structure and service elements left proudly on view echoed Paxton's Crystal Palace and the engineering glories of the railway age. It filtered down to furniture, such as the 1981 Transit airport seating by Rodney Kinsman and Foster's own Nomos Desking System (1986) for the Italian manufacturer Tecno.

This modernist style was not to everybody's taste. In the most controversial episode in postwar British architecture, in 1984 the Prince of Wales spoke out against a proposed modern extension to the National Gallery in Trafalgar Square, saying it would be like "a monstrous carbuncle on the face of a much-loved and elegant friend." The scheme was dropped. Advised by leading postmodernists such as Léon Krier, he went on to articulate his own "vision of Britain" on national television and set about building a traditionalist model town on crown land.

The image of the 1979 general election that swept Margaret Thatcher into power was created by the foremost homegrown advertising agency, Saatchi and Saatchi. It showed an endless line of unemployed with the ambiguous slogan "Labour isn't working." Saatchi and Saatchi was one of a new breed of advertising agency—self-consciously creative rather than driven by the hard sell like its American counterparts. Rivalry naturally grew up between design and advertising as to what extent any client's success could be attributed to their respective efforts. This led to a certain amount of knowing cross-referencing. In one beer advertisement of the time, for example, a trendy loft dweller is seen ejecting his old furniture and installing Ron Arad's Rover Chair (1981).

The Labour government had failed, and Thatcher set about showing that the national labor force, too, was not in her view up to scratch. Her decade-long battle with the labor unions largely finished off what remained of British manufacturing. The question now was whether a creative and service sector, driven by companies like Saatchi and Saatchi and other forms of creative agency (animation, television production, computer games) as well as banks and finance houses, could sustain a national economy in its place. Could the country prosper making Rover Chairs and Rover Chair commercials in place of Rover cars?

Thatcher's policies certainly benefited most designers. Not only were they mostly the sort of self-employed entrepreneurs she admired, but also they found themselves much in demand by a slew of new corporate clients for whom a swanky appearance was what mattered most in the thrusting marketplace, if not to their customers then to a new constituency, their shareholders. Several of the larger design consultancies were sufficiently caught up in the spirit of the age to seek a listing on the stock exchange. Companies such as British Airways and British Telecom were privatized. In some cases, they celebrated their newfound freedom by stamping a distinctive new corporate identity and learning new marketing and service skills. But in others, such as that of the former national railways, a well-developed ethos of design management disintegrated more or less overnight.

Napoleon had called Great Britain a nation of shopkeepers, and now under Mrs. Thatcher, the daughter of a grocer, the epithet seemed more apt than ever. Retail design impresarios such as Michael Peters and Rodney Fitch transformed the British high street from end to end, working for clients from supermarkets to boutiques. The revolution took in established names such as Boots the Chemist and newsagent W. H. Smith and newcomers such as the Body Shop, Laura Ashley, and Next. At the bespoke end of the trade, Norman Foster designed for

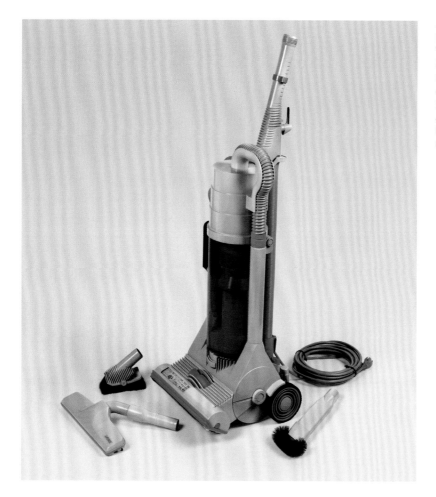

James Dyson (British, born 1947). G-force Vacuum Cleaner. 1986. Plastic and electro-mechanical parts, dimensions variable. Manufacturer: APEX, Japan. Victoria and Albert Museum, London. Given by Dyson

Katharine Hamnett and Nigel Coates for Jasper Conran. A new generation of "high tech" stylists, led by Future Systems and Eva Jiricna, also cut their teeth designing fashion stores. Though British-designed, many of the clothes sold in these shops were made abroad, a situation from time to time noisily decried by their creators.

The story of James Dyson and his "cyclone" vacuum cleaner is hardly typical of the experience of British design and manufacturing in the 1980s, but it does provide an instructive series of tableaux in design-led innovation. Dyson's innovation was technical: to use a vortex of air to collect dust in an easily emptied bucket in place of conventional bags that quickly clogged. After making thousands of prototypes, he had a viable design, but the only interest from British manufacturers was in suppressing his invention. He turned to Japan, where the cleaner was put into production with a price tag of three thousand dollars and aggressive styling to match. Dyson was able to use profits from the sales of this cult object to modify the design for manufacture on a larger scale and sale at a more modest price. He built his own factory in Britain to make the mass-market product, but price factors and quality-control issues soon forced him to move production offshore to Malaysia. Dyson's vacuum cleaners are not the neatly resolved forms that would have counted for standout industrial design a decade or two before. Inspired (like Richard Rogers's architecture) by the British engineering tradition, they delight in the exposure of corrugated hoses, clip-together components, and an expressionist style that owes something to comic-book spacecraft. But this exuberance should not obscure the engineering truth that they work better than the competition.

The advent of the European single market in 1992 gave citizens of member states the right to work anywhere in the European Union, but some of the best British designers had long ago decided to take advantage of opportunities overseas. Domestic automotive manufacturing may have declined, but British designers have been studio heads at leading Continental companies, such as Audi. David Lewis has, since 1965, guided the distinctive house style of the Danish hi-fi manufacturer Bang and Olufsen, and Perry King, George Sowden, and James Irvine are among those to have made their mark in Milan. British designers have been influential in bringing a more sophisticated, European look to the design of computers emerging from California, a trend that has culminated (for now) in the astonishing success of the British-born designer Jonathan Ive at Apple Computer in establishing the paradigm-shifting style of successive generations of Macintosh computers, the iPod MP3 player, and the iPhone.

Conversely, many foreign designers have decided to make Great Britain (in almost all cases, London) their professional home, and their arrival has once again done much to invigorate the local scene. Among those who have made this journey are Ron Arad (from Israel), Zaha Hadid (Iraq), Daniel Weil (Argentina), Danny Lane (USA), Hussein Chalayan (Cyprus), Shin and Tomoko Azumi (Japan), Tord Boontje (the Netherlands), and others. The draw for them is clearly not the abundance of design-literate manufacturers. Rather it is the boundary-breaking possibilities offered by the capital, with its global reputation in contemporary art and media and some of the world's best art schools.

This traffic has made British design as pluralistic and open to ideas as it has ever been. Three leading furniture designers show both the range that is possible and the common threads. Tom Dixon has abandoned some of the more baroque excesses that characterize his work of the 1980s (of which the sinuous steel frame woven with rush of the S Chair, of 1991, is but a mild example). He became the creative director at Habitat in 1997 but has continued to develop his own designs independently. His 1996 polypropylene Lamp Jack, a simple geometric form that doubles as lamp and stool, is typical of the quiet refinement and acceptance of modern materials that have entered his work. Ross Lovegrove's Figure of Eight Chair (1993) is considered an update of Arne Jacobsen's famous Ant Chair of 1952, but it has a forthright simplicity that makes it a natural successor to Robin Day's still-ubiquitous 1962 Polyprop stacking chair. Jasper Morrison has always exhibited a somewhat ascetic streak, and it now seems right for the times. His 1984 Wingnut Chair is folded from thick cardboard, while the Plywood Chair of 1988, a deceptively simple design solely assembled from cut thin plywood, carries the folk memory of postwar austerity but on second glance reveals a stark beauty. Morrison has continued in this vein with his 2007 Cork Chair and Stools, made entirely of recycled bottle corks, for the Swiss manufacturer Vitra. None of these designs is exactly minimal or utilitarian, but there is no fat here, only a relaxed assuredness in the handling of materials and form.

Long-awaited by some, the electoral victory of the New Labour party in 1997 brought a fresh wave of optimism to a design industry that at last seemed to have found its bearings, with London's creative talent well connected to receptive clients and manufacturers in Europe and Asia. A great celebration of national creativity was planned for the turn of the century in a massive Thames-side fabric structure, the Millennium Dome, and £733 million was spent. The fact that the result was a huge disaster—exhibits ranged from the vacuous to the pompous with little in between—can be attributed to politicians' congenital inability to trust creative decision making.

A series of architectural commissions, also completed around the turn of the millennium, resulted in more-popular additions to the British cultural landscape. Though instigated by the outgoing Conservative government and funded by the national lottery it had introduced, these schemes—the French, more used to such things, would call them grands projets—appeared to everyone to belong to the New Labour era, dubbed "Cool Britannia." Best of the more than two hundred funded schemes is the Eden Project, a cluster of geodesic domes in a disused Cornish quarry housing a botanic garden, designed by Nicholas Grimshaw and Partners. Other projects, such as the National Centre for Popular Music in Sheffield, by Nigel Coates, were unsuccessful in attracting the interest of the public and eventually failed. The overall consequence of all this largesse, however, is that Britain has found itself, slightly to its surprise, comfortable at last with late modernism. These new landmarks—funded by the people, for the people—are the evidence. Whether this newly relaxed attitude stems from true conviction or is a flash in the pan, as the acceptance of modern design arguably was at the time of the Festival of Britain, remains to be seen.

One of New Labour's first actions was to bring more devolved government to Britain's regions. The Scots celebrated with an extravagant—and hugely over-budget—parliament building by the Catalan architect Enric Miralles (1998–2004). It remains one of very few examples of Britain returning the favor of patronage so visibly extended to its architects overseas.

In design, as in many areas of life, Britain's regions struggle to establish a distinctive identity. Scottish design consultancies, such as McIlroy Coates (later EH6) in Edinburgh and Graven Images in Glasgow, have succeeded in this. But it is to their respective cities as much as to

Grimshaw Architects LLP (UK, established 1980). Eden Project, Cornwall, England. 2001–02. Aerial view, 2006

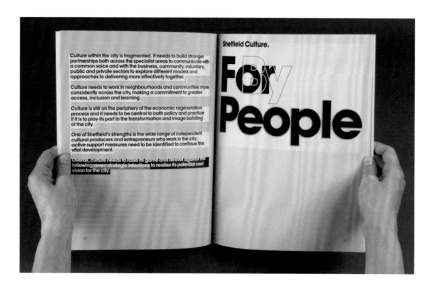

The Designers Republic (UK, established 1986). Ian Anderson (British, born 1961). Design for *Sheffield Cultural Strategy* (Sheffield, England: Sheffield Town Council, 2006). Offset lithography, softcover book, closed: 9⁷⁄₁₆ x 11⅝" (24 x 29.5 cm)

the Scottish nation that their work relates, as Janice Kirkpatrick of Graven Images has explained: "Design can go some way towards revealing cultural differences, giving us an interpretation of why Glasgow is different from other cities and how we can help reveal and enhance these differences for human and economic benefit."[7]

Elsewhere, too, it is cities that provide the focus. At the end of the 1970s, Manchester was where graphic designers Peter Saville and Malcolm Garrett and interior designer Ben Kelly created the aesthetic, driven by avant-garde European typography, at Factory Records and the Haçienda Club. More recently, The Designers Republic, a self-consciously rebellious design consultancy based in Sheffield, has specialized in bold, colorful graphics for a range of mostly street-credible clients—Swiss modernism for the rave generation. It is no coincidence that these cities have been the nurseries of some of Britain's biggest bands of recent years.

Perhaps as a consequence of the long-term decline of British manufacturing, profound shifts occurred toward the end of the twentieth century in the nature of product design. As more products have come to incorporate digital technology, traditional ergonomics has been supplanted by an emphasis on the design of easily usable computer interfaces. British designers have risen well to this new challenge. Bill Moggridge of IDEO was among the first to use the term "interaction

Anthony Dunne
(British, born 1964).
Fiona Raby (British,
born 1963). Faraday
Chair, prototype. 1996.
Perspex and steel,
27½ x 27½ x 48⅜"
(70 x 70 x 123 cm)

design" to describe this new area of activity, in which designers were clearly the professionals best qualified to make improvements. In 1989 the Royal College of Art opened a graduate department called Computer-Related Design (now Design Interactions), for which Moggridge has been a visiting professor, and gradually enlarged the scope of this growing specialty. As this new discipline has matured, its attitude toward technology has undergone a subtle but significant shift. Designers in this area began by simply helping manufacturers make their feature-laden digital devices more usable. Now they are in a position to advise that disguising, or even reducing, the technological complexity of the interface may be the best course. This has led to the reappearance of analog or quasi-analog features such as rotary dials in place of buttons and the introduction of a greater element of play. An example of the trend is Philip Worthington's Shadow Monsters (2004), a screen-based interface where the computer, reacting to hand movements, adds its own embellishments to shadow puppets created by the user.

This new perspective may prove to affect the entire demeanor of design toward technology and society. More than ever, the responsible role of the designer is to be no longer the loyal enabler for a client but a critic of society. The implicit criticism of computer technology in such projects easily extends to a broader critique of our technological lives, which in turn brings the social function of the designer closer to that of the artist. Tony Dunne and Fiona Raby have been among the designers behind this development, first through polemical projects and more lately through their teaching. The Huggable Atomic Mushroom (2005), designed with Michael Anastassiades, and the Faraday Chair (1998), a minimal metal-screened cabinet in which one may shelter from electromagnetic radiation, address our technological fears with sardonic wit and a good dose of the placebo effect. Dunne is Head of Design Interactions at the Royal College of Art. Recent graduates from his course have sought critical engagement with new fields, such as genetic engineering and nanotechnology. For example, Revital Cohen's project Life Support, of 2008, imagines symbiotic animal companions used in place of impersonal life-support machines. It is clear that these are not prototypes for manufacture, and it is equally clear that designers in this field are not presenting quite the account of themselves that conventional industrial employers would wish to see. Where these graduates will find their place professionally is for the moment a bit of a mystery.

Not far from the minds of some of them is probably the success story of the latest generation of British Conceptual artists, mainly graduates of London's Goldsmiths College taught by the American Michael Craig-Martin in the 1990s. The work of these young British artists (YBAs) was bought, displayed, and sold (at huge profits) by the former adman Charles Saatchi. In terms of influence, however, it is a slightly older generation of artists concerned with material culture, such as Richard Wentworth and Cornelia Parker, who most inspire the work of these new designers. A key crossover figure in this is Ron Arad, whose furniture has long commanded art-house prices and whose commitment to the "one-off"—as his studio was long-ago named—is unquestioned. Arad, influential as a teacher as well as a practitioner, has been among the first to embrace rapid manufacturing (whereby an object is fabricated directly from a reservoir of resin or other raw material, based on a digital design). Rapid manufacturing promises to blur once and for all the distinction between the unique creation of the artist and the design executed for mass manufacture. Reconsider Adam Smith's pin: once designed in the knowledge that it would have to be manufactured in numerous stages, it can now, in principle, be made in a single pass, by a machine (quite possibly located in the home) that is fed instructions by a user who is designer, maker, and consumer in one.

But is this any basis for a national economy? The global financial meltdown of 2008 has shown that Britain is out on a limb with its postindustrial experiment. Can the country, as its designers have done much to suggest, create economic success in a virtual world? Or must a way be found to re-establish a manufacturing industry? If the latter is the case, then it is possible that designers are once again prescient in their exploration of rapid manufacturing, a technology that will put them at the center of creation and banish to the margins the faint-hearted company men who have for so long failed to recognize the talent on their doorstep.

This is what British design has been and is. But what of its future? (And what, indeed, is the future of recognizably national design from anywhere?) The country is uneasily poised.
A national style of design is not an irrelevance in Great Britain as, for different reasons, it probably is for the United States or Germany or Korea. But nor is it quite the tempting comfort blanket that it is for countries like Denmark that imagine they can prosper by marketing an *appellation contrôlée* version of the national style. Britain is too large and embodies too many contradictions to attempt this. Instead, the country must resist its tendency toward insularity. It is when its designers look outward that interesting things begin to happen.

The national tendency to compromise is a good one—it increases the likelihood of manufacturability and consumer acceptance, which design will always depend on. But the possibilities from which compromise is to be reached must come from somewhere. It is the demands and the benefits of the Industrial Revolution, of colonial and economic crisis, of international immigration, and of globalization that have repeatedly led British designers to renegotiate the balance between craft and technology, delight and practicality, and puritanism and excess and still, more often than not, meet somewhere in the middle.

In 2012 London will host the Olympic Games. What should the world expect? The tug of war has continued between the forces of revolution and compromise, and some exciting plans have fallen by the wayside due to political niggardliness or their creators' overweening ambition. Still, the thing will probably be competently executed. The design will be mostly unimaginative but exhibit occasional flashes of genius, and the British people will be more excited by it all than they let on.

1. William Lethaby, reply to an address, Oxford University, January 18, 1922. Quoted in Fiona MacCarthy, *William Morris* (London: Faber and Faber, 1994), p. 358.
2. Liz McQuiston, *Graphic Agitation* (London: Phaidon, 1993), p. 19.
3. Adrian Forty makes this criticism in *Objects of Desire* (New York: Pantheon, 1986), p. 237.
4. Nikolaus Pevsner, 1942. Quoted in Oliver Green and Jeremy Rewse-Davies, *Designed for London: 150 Years of Transport Design* (London: Laurence King, 1995), p. 71.
5. Jeremy Myerson, *Gordon Russell* (London: Design Council, 1992), p. 85.
6. Frederique Huygen uses these adjectives in her astute study *British Design: Image and Identity* (London/Rotterdam: Thames and Hudson/Museum Boijmans Van Beuningen, 1989), p. 19.
7. Janice Kirkpatrick, "City Culture Pays," in Jeremy Myerson, ed., *Design Renaissance* (Horsham, England: Open Eye, 1994), pp. 82–83.

1700_1929

Josiah Wedgwood

1730–1795

Coffee Pot and Lid, 1768
Black basalt with glazed interior,
6⅝ x 7¼ x 4¼" (16.8 x 18.4 x 10.8 cm)
Manufacturer: Josiah Wedgwood
& Sons Ltd., Stoke-on-Trent, England
Gift of Josiah Wedgwood & Sons Inc.
of America, 1954

Josiah Wedgwood

1730–1795

Demitasse Cup and Saucer, 1768
Black basalt with glazed interior,
demitasse cup: 2¼ x 3¼″ (5.7 x 8.3 cm),
saucer: ⅞ x 5¾″ diam. (2.2 x 14.6 cm)
Manufacturer: Josiah Wedgwood
& Sons Ltd., Stoke-on-Trent, England
Gift of Josiah Wedgwood & Sons Inc.
of America, 1954

Josiah Wedgwood

1730–1795

Teacup and Saucer, 1768
Black basalt with glazed interior,
teacup: 2¼ x 2³⁄₁₆″ (5.7 x 5.5 cm),
saucer: ¾ x 4⅜″ diam. (1.9 x 11.1 cm)
Manufacturer: Josiah Wedgwood
& Sons Ltd., Stoke-on-Trent, England
Gift of Josiah Wedgwood & Sons Inc.
of America, 1954

Charles Rennie Mackintosh

1868–1928

Side Chair, 1897
Oak and silk, 54⅜ x 20 x 18″
(138.1 x 50.8 x 45.7 cm)
Gift of the Glasgow School of Art,
1958

Harry J. Powell

1855–1922

Wine Glass, 1880–1914
Glass, 4⅞ x 2¾" diam.
(12.4 x 7 cm)
Manufacturer: Whitefriars
Glassworks, London
Mrs. Armand P. Bartos Purchase
Fund, 1996

Philip Webb

1831–1915

Finger Bowls, c. 1880
Vaseline glass, each: 3 x 5¾″ (7.6 x 14.6 cm) (irreg.)
Manufacturer: J. Powell & Sons, London
Dorothy Cullman Purchase Fund, 2002

Christopher Dresser

1834–1904

Claret Pitcher, c. 1880
Glass, silver plate, and ebony,
16⅝ x 5¼ x 4"
(42.2 x 13.3 x 10.2 cm)
Manufacturer: Hukin & Heath,
Birmingham
Gift of Blanchette Hooker
Rockefeller, 1977

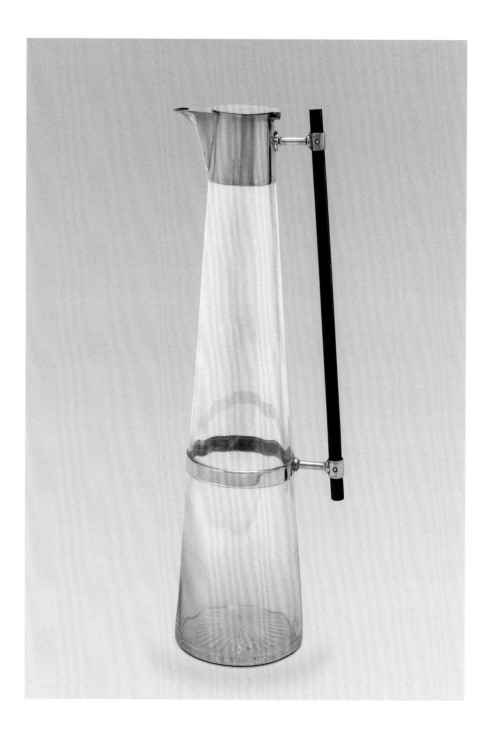

Christopher Dresser

1834–1904

Watering Can, c. 1876
Painted tinned-iron,
12⅝″ x 9⅞″ x 7¼″
(32 x 25.1 x 18.4 cm)
Manufacturer: Richard Perry,
Son & Company, Wolverhampton,
England
Gift of Paul F. Walter, 1993

56

Charles Rennie Mackintosh

1868–1928

The Scottish Musical Review, 1896
Lithograph, 8′1″ x 37″
(246.4 x 94 cm)
Printer: Banks & Co., Edinburgh
and Glasgow
Acquired by barter from
the University of Glasgow, 1960

Charles Rennie Mackintosh

1868–1928

Fish Knife and Fork, c. 1900
Silver-plated nickel,
fork: 9⅛ x 1¼″ (23.2 x 3.2 cm),
knife: 8⅞ x 1⅛″ (22.5 x 2.9 cm)
Gift of the University of Glasgow,
1957

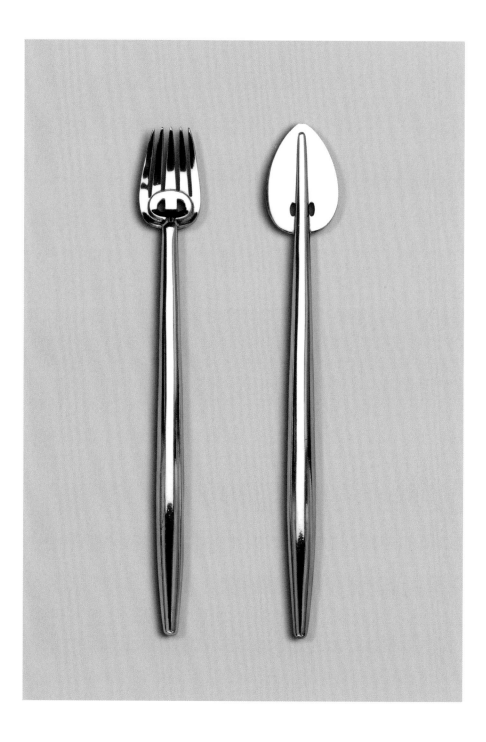

Charles Robert Ashbee

1863–1942

Silver Loop-handled Dish, 1901
Silver and lapis lazuli,
2¹¹⁄₁₆ x 7¹³⁄₁₆ x 4⅜″ (6.8 x 19.8 x 11.1 cm)
Manufacturer: Guild of Handicraft Ltd., London
Estée and Joseph Lauder Design Fund, 1977

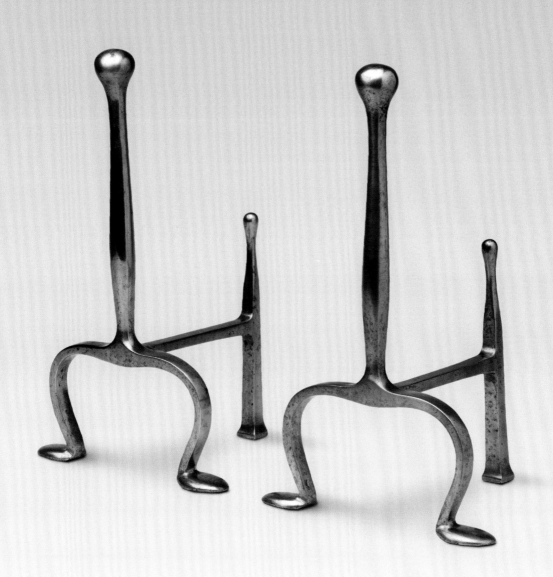

Archibald Knox

1864–1933

Jewel Box, c. 1900
Silver, mother-of-pearl, turquoise,
and enamel, 4 x 11½ x 6½"
(10.2 x 29.2 x 16.5 cm)
Manufactured by H. C. Craythorne
for Liberty & Co. Ltd., Birmingham
Gift of the family of Blanchette
Hooker Rockefeller, 1949

Eileen Gray

1879–1976

Adjustable Table, 1927
Chrome-plated tubular steel,
sheet steel, and glass,
min. h. 21¼″ (54 cm), max. h. 36½″
(93 cm) x 20″ diam. (50.8 cm)
Manufacturer: Aram Designs Ltd.,
London
Philip Johnson Fund and Aram
Designs Ltd., London, 1977

E. McKnight Kauffer

American, 1890–1954

Read "Cricketer" in the Manchester Guardian, 1923
Lithograph, 30 x 19¹³⁄₁₆"
(76.2 x 50.3 cm)
Gift of the designer, 1939

Tom Purvis

1889–1959

East Coast Resorts, 1925
Lithograph, 39½ x 50″
(100.3 x 127 cm)
Given anonymously, 1968

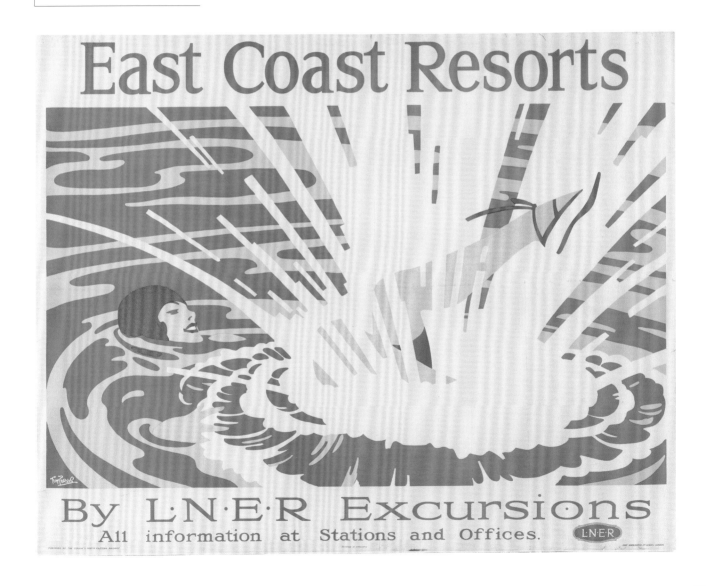

IN THE **BLACK**OUT

the_thirties_forties_and_fifties

19**30**_19**59**

Marcel Breuer

1902–1981

Chaise Longue, 1935–36
Bent birch wood and upholstered
cushion, 32 x 58 x 24″
(81.3 x 147.3 x 61 cm)
Manufacturer: The Isokon
Furniture Company, London
Purchase Fund, 1942

Eileen Gray

1870–1976

Tube Lamp, c. 1930s
Chromed metal and incandescent
tube, 36 x 9⅞" diam.
(91.5 x 25.1 cm)
Estée and Joseph Lauder Design
Fund, 1980

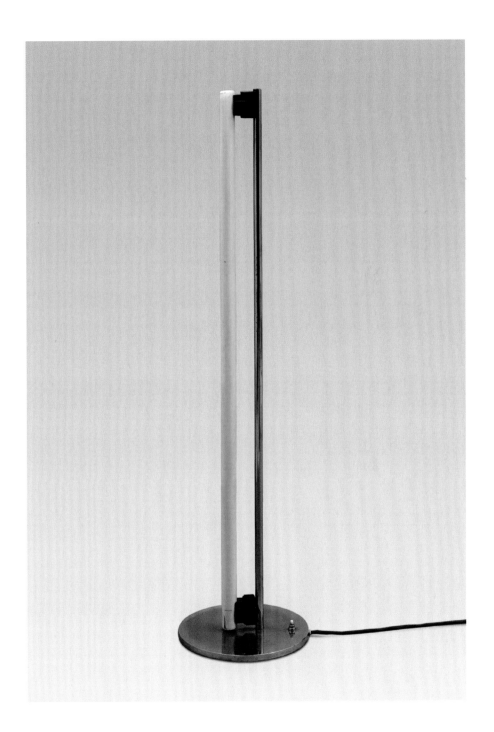

Edgar Ainsworth

1906–1975

*Everywhere You Go, You Can
Be Sure of Shell. Gordale Scar
—The Craven Fault, Yorks,* 1934
Lithograph, 30 x 45"
(76.2 x 114.3 cm)
Gift of Shell-Mex BP, 1937

Paul Nash

1889–1946

You Can Be Sure of Shell.
Footballers Prefer Shell, 1935
Lithograph, 30¹/₁₆ x 42¾"
(76.4 x 108.6 cm)
Gift of G. E. Kidder Smith, 1943

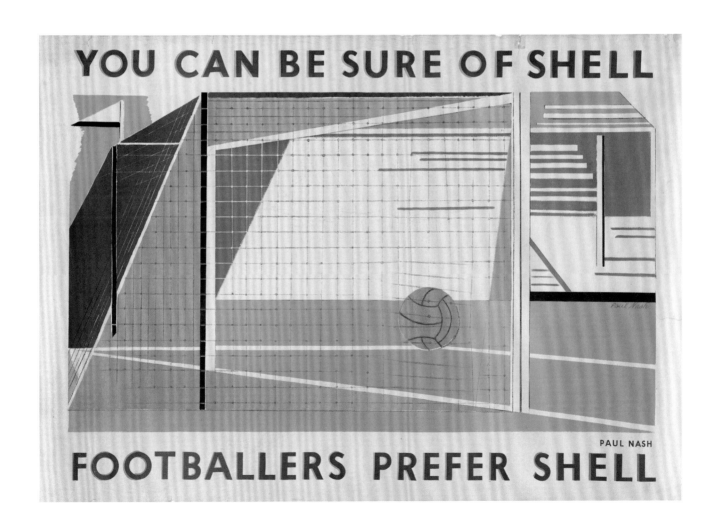

J. S. Anderson

Dates unknown

You Can Be Sure of Shell.
Motorists Prefer Shell, 1935
Offset lithograph, 30 x 44¾"
(76.2 x 113.6 cm)
Purchase, 1968

J. S. Anderson

Dates unknown

To Visit Britain's Landmarks,
You Can Be Sure of Shell,
c. 1939–43
Lithograph, 30 x 45"
(76.2 x 114.3 cm)
Gift of G. E. Kidder Smith, 1943

Gerald Summers

1899–1967

Lounge Chair, 1934
Bent birch plywood with pigmented lacquer,
29⅝ x 35 x 23½″ (75.2 x 88.9 x 59.7 cm)
Manufacturer: Makers of Simple Furniture
Ltd., London
Barbara Jakobson Purchase Fund
and Peter Norton Purchase Fund and
Gift of Robert and Joyce Menschel, 2000

Ben Nicholson

1894–1982

Imperial Airways, 1935
Lithograph, 38¾ x 23⅝"
(98.4 x 60 cm)
Gift of the designer, 2000

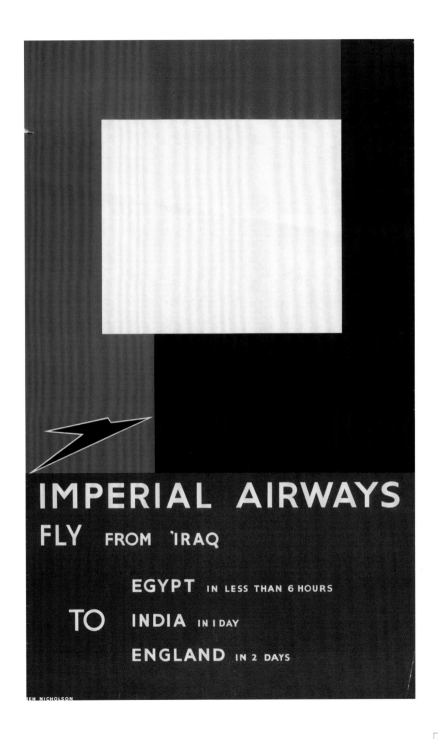

Edward Bawden

1903–1989

Kew Gardens. Kew Gardens Station, 1937
Lithograph, 40 x 25″
(101.6 x 63.5 cm)
Printer: Curwen Press, London
Gift of Philip Johnson, 1950

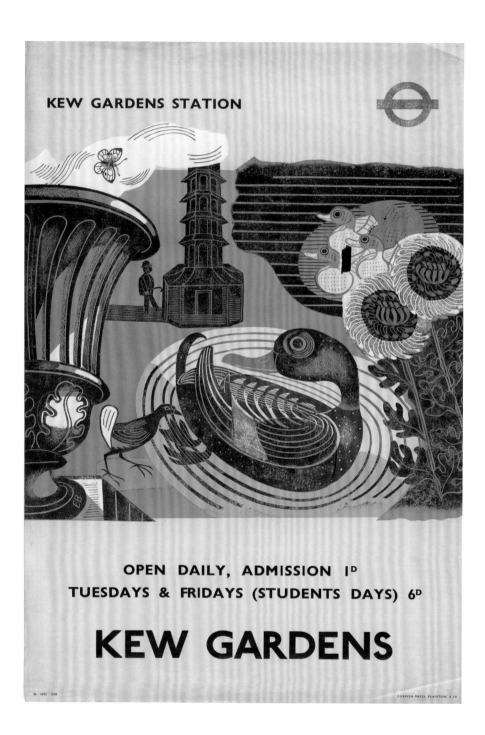

Thomas Eckersley

1914–1997

Time to Change to Winter Shell, 1938
Offset lithograph, 29⅞ x 45"
(75.8 x 114.3 cm)
Gift of The Lauder Foundation,
Leonard and Evelyn Lauder Fund, 1983

Abram Games

1914–1996

Talk Kills, 1942
Lithograph, 29 x 19″
(73.7 x 48.3 cm)
Gift of the artist, 1945

Frederick H. K. Henrion

1914–1990

*Action Stations. Saving Is
Everybody's War Job*, 1939–45
Photolithograph, 29⅛ x 35½″
(74 x 90.2 cm)
Gift of the artist, 1945

Abram Games

1914–1996

Grow Your Own Food, 1942
Photolithograph,
29⅜ x 19³⁄₁₆″ (74.7 x 48.6 cm)
Printer: J. Weiner Ltd., London
Gift of Mrs. John Carter, 1943

Jan Le Witt

Born Poland. 1907–1991

George Him

Born Poland. 1900–1982

Food Needs Transport. Don't Waste It, 1944
Lithograph, 19⅝ x 29⅜" (49.8 x 74.6 cm)
Printer: Geo. Gibbons Ltd., Leicester, England
Gift of the artists, 1947

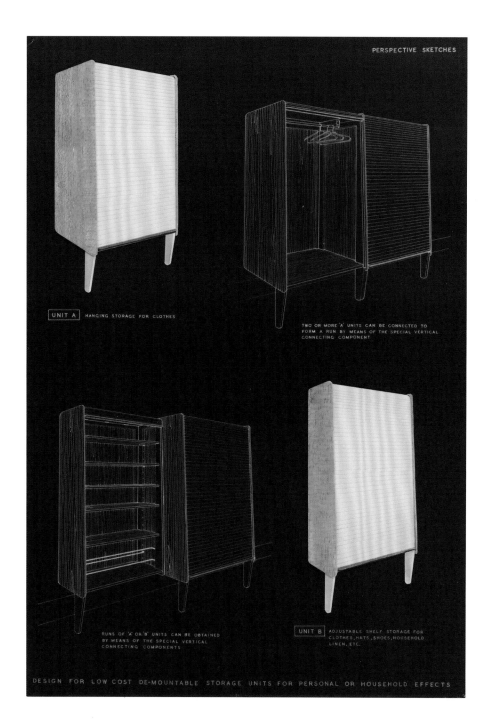

PERSPECTIVE SKETCHES

UNIT A | HANGING STORAGE FOR CLOTHES

TWO OR MORE 'A' UNITS CAN BE CONNECTED TO
FORM A RUN BY MEANS OF THE SPECIAL VERTICAL
CONNECTING COMPONENT

RUNS OF 'A' OR 'B' UNITS CAN BE OBTAINED
BY MEANS OF THE SPECIAL VERTICAL
CONNECTING COMPONENTS

UNIT B | ADJUSTABLE SHELF STORAGE FOR
CLOTHES,HATS,SHOES,HOUSEHOLD
LINEN, ETC.

DESIGN FOR LOW COST DE-MOUNTABLE STORAGE UNITS FOR PERSONAL OR HOUSEHOLD EFFECTS

Ernest Race

1913–1964

*Entry Panel for the MoMA
International Competition for
Low-Cost Furniture Design,* c. 1940
Ink and collage on panel,
29¾ x 20″ (75.6 x 50.8 cm)
Gift of the designer, 2008

Zero (Hans Schleger)

1898–1976

*In the Blackout, Wear or Carry
Something White,* 1943
Lithograph, 25 x 20″
(63.5 x 50.8 cm)
Gift of the London Passenger
Transport Board, 1945

IN THE **BLACKOUT**

Wear or carry
something white

Ernest Race

1913–1964

BA Side Chair, 1945
Enameled cast aluminum, plywood,
and cotton, 30$\frac{1}{2}$ x 17$\frac{1}{2}$ x 19$\frac{1}{4}$"
(77.5 x 44.5 x 48.9 cm)
Manufacturer: Race Furniture Ltd.,
Sheerness, England
Gift of Waldron Associates, 1952

Lucie Rie

1902–1995

Bowl, c. 1953
Glazed porcelain, 3¾ x 5½″ diam.
(9.5 x 14 cm)
Purchase Fund, 1954

Michael Rabinowitz

Dates unknown

Radio, 1946
Chrome-plated steel and plastic,
19¼ x 12½″ diam. (50.2 x 31.8 cm)
Department Purchase Fund, 1989

Philip Vincent
1908–1982
Phil Irving
1903–1992

Vincent-HRD Series C Black Shadow
Motorcycle, 1949
Painted and chromium-plated metal, polished
aluminum, rubber, leather seat, and rubber tires,
45½″x 7′3½″ x 28¼″ (115.6 x 222.3 x 71.8 cm)
Manufacturer: Vincent-HRD Company Ltd.,
Stevenage, England
Gift of James Gubelmann and the Tropical
Gangsters Motorcycle Racing Team, 2003

Abram Games

1914–1996

Jersey, 1951
Lithograph, 40¼ x 25"
(102.2 x 63.5 cm)
Gift of the artist, 1953

Robert Welch

1929–2009

Nutcracker, 1958
Stainless steel, 6 x 1⁷⁄₈″
(15.3 x 4.8 cm)
Manufacturer: J. & J. Wiggin Ltd.,
Old Hall Works, London
Gift of the manufacturer, 1960

Fly
the Tube

the_sixties_seventies_and_eighties

1960_1989

O. F. Maclaren

1906–1978

Baby Stroller, 1966
Aluminum alloy tubing and saran
polythene fabric,
folded: 41 x 6½ x 6½″
(104.2 x 16.5 x 16.5 cm),
open: 36¾ x 14⅝ x 35″
(93.3 x 37.1 x 88.9 cm)
Manufacturer: Andrew Maclaren
Ltd., England
Gift of the designer, 1970

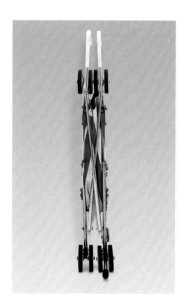

Russell Manoy

Born 1945

Mug and Plate, 1966–67
Melamine, mug: 5 x 2¾" diam. (12.7 x 7 cm)
plate: 1¾ x 11 x 7" (4.4 x 27.9 x 17.8 cm),
Manufacturer: Antiference Ltd., Aylesbury, England
Friends of the Department Fund, 1988

REVOLVER

Klaus Voormann

Born 1938

Cover for The Beatles' LP Record
Revolver, 1966
Lithograph, 12⅜ x 12⅜"
(31.4 x 31.4 cm)
Gift of Christian Larsen, 2008

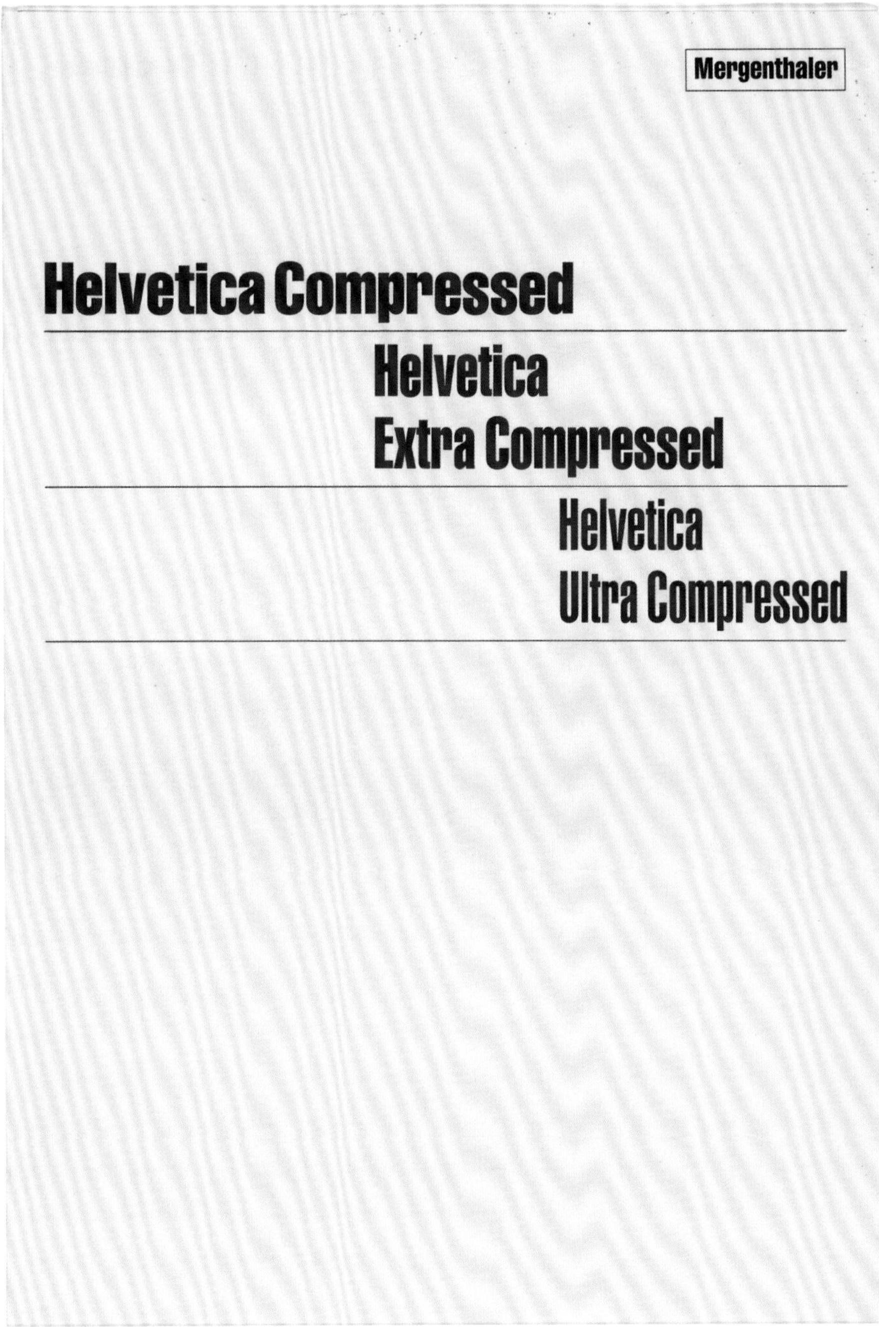

Mergenthaler

Helvetica Compressed

Helvetica
Extra Compressed

Helvetica
Ultra Compressed

Mergenthaler Linotype Company

Established 1886

Matthew Carter

Born 1937

Hans-Jürg Hunziker

Born 1938

Helvetica Compressed Type
Specimen, 1966
Lithograph, 11⅞ x 8½"
(30.2 x 21.6 cm)
Gift of Matthew Carter, 2008

Charles Currey

1890–1973

Sailing Knife and Marlinespike
(model 468), c. 1960
Stainless steel,
closed: 1⁷⁄₁₆ x 3⁵⁄₈″ (3.7 x 9.2 cm)
Manufacturer: Blyde & Co. Ltd.,
England
Gift of The Crow's Nest, 1962

Hans Coper

1920–1981

Vase, c. 1963
Stoneware, 6 x 4⅛″ diam.
(15.3 x 10.5 cm)
Given anonymously, 2001

Arthur Sanderson & Sons Ltd.

Dates unknown

William Morris

1834–1896

May Morris

1862–1938

Owen Jones

1809–1874

William Burges

1827–1881

John Henry Dearle

1860–1932

Wallpaper Sample, c. 1965
Block print on paper,
8 x 5″ (20.3 x 12.7 cm)
Purchase, 2008

Peter Gee

1932–2005

Color Image, 1966
Silkscreen on metallic paper,
21¾ x 21½″ (55.3 x 54.6 cm)
Gift of the designer, 1966

Color Image

The Basic Research Unit
Summer Workshop 1966
June 7th through June 16th
Full details and application form
Box 2G 506 West Broadway
New York 12, NY

Workshop conducted by
Peter Gee Instructor of Color
Harvard University

Alan Fletcher

1931–2006

Pirelli, 1961
Photolithograph,
36 x 14⅞" (91.4 x 37.7 cm)
Gift of the designer, 1961

Eduardo Paolozzi

1924–2005

Universal Electronic Vacuum 1967,
1967
Silkscreen and offset lithograph,
34⅜ x 24¼″ (87.3 x 61.6 cm)
Gift of the Pace Gallery, 1968

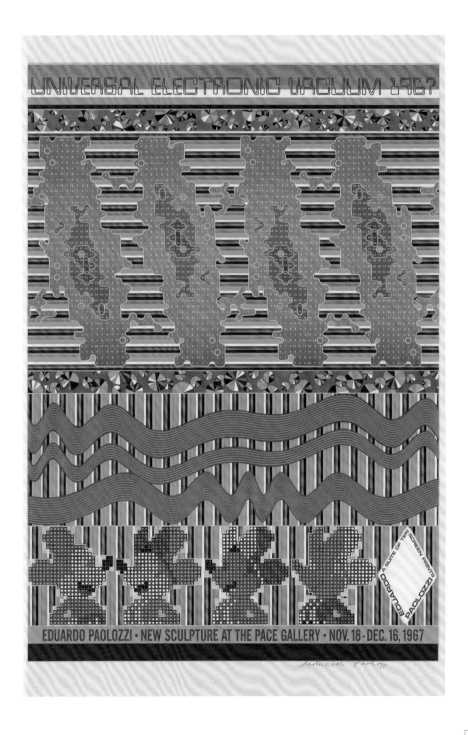

Michael English

1941–2009

Love Festival, 1967
Silkscreen, 29⅞ x 40″
(75.8 x 101.6 cm)
Gift of P. Reyner Banham, 1968

Brian J. Watson

Born 1949

Fly the Tube, 1977
Offset lithograph, 39⅜ x 24⅝"
(100 x 62.5 cm)
Printer: Foote, Cone and Belding
Advertising Ltd., USA
Gift of the London Transport
Enquiry Office, 1981

Martin Roberts

Born 1943

Conran Associates

Established 1955

Input 14 Ice Bucket, 1973
ABS polymer,
6¾ x 7″ diam. (17.1 x 17.8 cm)
Manufacturer: Crayonne Ltd.,
Sudbury-on-Thames, England
Gift of Conran's, 1977

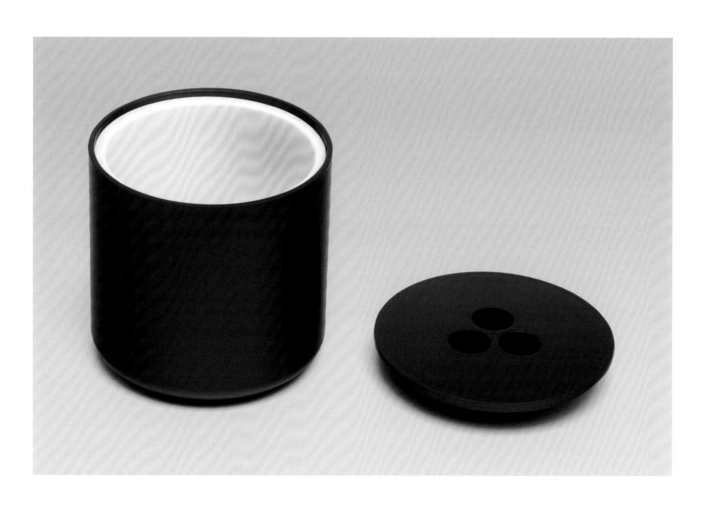

John Pemberton

Born 1948

Sovereign Calculators, 1975
Metal casing and plastic,
each: ½ x 1½ x 5⅝"
(1.3 x 3.8 x 14.3 cm)
Manufacturer: Sinclair Radionics
Ltd., Huntingdon, England
Gift of Sinclair Radionics, 1978

Bill Moggridge
Born 1943
Stephen Hobson
Born 1942
Glenn Edens
Born 1952

Compass Portable Computer, 1981
Die-cast magnesium casing and
injection-molded plastic, closed:
2¹/₁₆ x 11½ x 15″
(5.2 x 29.2 x 38.1 cm)
Manufacturer: Grid Systems Corp.,
Mountain View, California
Gift of the manufacturer, 1982

Daniel Weil

Born Argentina 1953

Bag Radio, 1981
Flexible PVC casing,
11½ x 8⁵⁄₁₆″ (29.2 x 21.1 cm)
Manufacturer: Parenthesis Ltd.,
London
Skidmore, Owings & Merrill Design
Collection Purchase Fund, 1983

Marcello Minale

1938–2000

Brian Tattersfield

Born 1936

British Airports at The Design Center, 1980
Offset lithograph,
29^{15}/$_{16}$ x 19^{15}/$_{16}$" (76 x 50.7 cm)
Gift of the artists, 1982

Nick Butler

Born 1942

Durabeam Flashlight, 1986–87
Plastic, 5¼ x 2⅜ x 1¼"
(13.3 x 6 x 3.2 cm),
open: h. 6⅜" (16.2 cm)
Manufacturer: Duracell International
Inc., Bethel, Connecticut
Anonymous gift, 2001

Alex Moulton

Born 1920

AM2 Bicycle, 1983
Painted steel tubing, metal,
leather, plastic, and rubber,
40 x 60 x 19"
(101.6 x 152.4 x 48.3 cm)
Barbara Jakobson Purchase Fund,
2003

Foster Associates

Established 1967

Nomos Dining Table, 1986
Chrome-plated steel and glass,
25¾″ x 7′2½″ x 39⅜″
(65.4 x 219.7 x 100 cm)
Manufacturer: Tecno SpA, Milan
Gift of the manufacturer, 1988

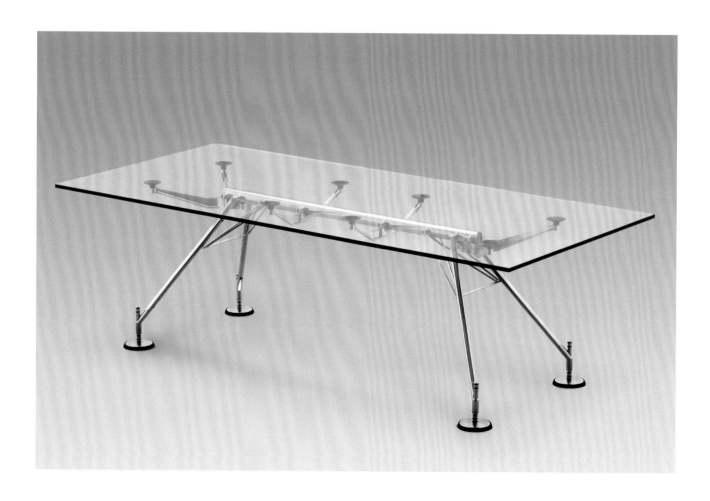

David Lewis

Born 1939

Video Cassette Recorder Model
Beocord VX 5000, 1989
Steel, aluminum, and plastic,
3¼ x 21¼ x 11¾″
(8.2 x 54 x 29.8 cm)
Manufacturer: Bang & Olufsen,
Struer, Denmark
Gift of Bang & Olufsen of America
Inc., 1991

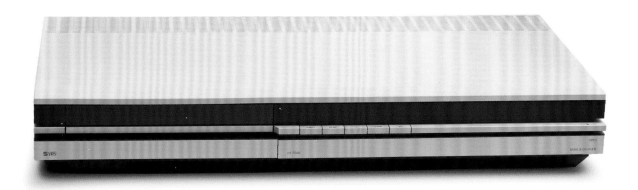

Tom Dixon

Born 1959

S Chair, 1991
Rush and steel,
40⅜ x 19¼ x 22⅞"
(102.6 x 48.9 x 58.1 cm)
Manufacturer: Cappellini SpA, Milan
Gift of the manufacturer, 1997

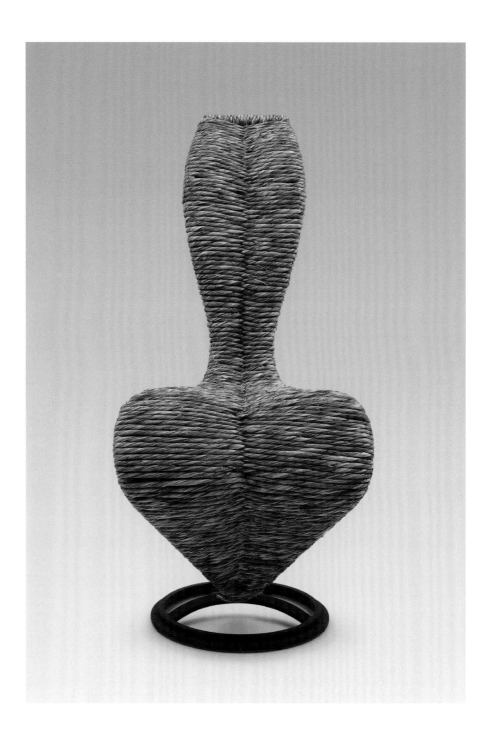

David Lewis

Born 1939

Beolab 6000 Loudspeakers, 1992
Aluminum case, Lycra, and plastic,
each: 43 x 4 x 3⅛″
(109.2 x 10.2 x 8 cm)
Manufacturer: Bang & Olufsen
of America, Mount Prospect, Illinois
Gift of the manufacturer, 1993

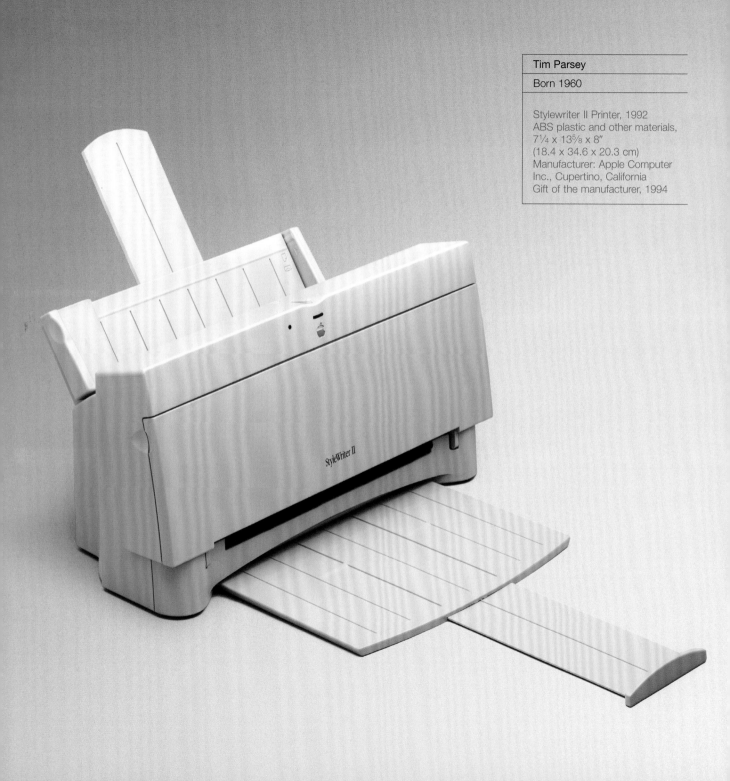

Tim Parsey

Born 1960

Stylewriter II Printer, 1992
ABS plastic and other materials,
7¼ x 13⅝ x 8″
(18.4 x 34.6 x 20.3 cm)
Manufacturer: Apple Computer
Inc., Cupertino, California
Gift of the manufacturer, 1994

StyleWriter II

James Dyson

Born 1947

Dual Cyclone Vacuum Cleaner
(model DC02), 1994–95
ABS, polycarbonate,
and polypropylene plastic,
19$\frac{11}{16}$ x 11$\frac{13}{16}$ x 18$\frac{1}{8}$"
(50 x 30 x 46 cm)
Manufacturer: Dyson Ltd., England
Gift of the manufacturer, 2003

Jasper Morrison

Born 1959

Bottle Storage Module, 1993
Injection-molded polypropylene
and anodized aluminum,
10¼ x 9 x 14³⁄₁₆″
(26 x 22.9 x 36 cm)
Manufacturer: Magis SpA, Italy
Gift of the manufacturer, 1998

Ross Lovegrove

Born 1958

Figure of Eight Chair, 1993
Polyurethane, stainless steel,
and nylon,
each: 33¼ x 19¾ x 21¾″
(84.5 x 50.2 x 55.2 cm)
Manufacturer: Cappellini SpA, Milan
Gift of the manufacturer, 1995

Tord Boontje

Born the Netherlands 1968

Emma Woffenden

Born 1962

"Transglass" Glassware, 1997
Recycled glass, left to right, vase two:
10 x 3" (25.4 x 7.6 cm), tumblers,
each: 3¼ x 2⅜" (8.3 x 6 cm), pitcher: 11 x 3"
(27.9 x 7.6 cm), double vase: 14½ x 3"
(36.8 x 7.6 cm), cup: 3¼ x 3¼" (8.3 x 8.3 cm),
carafe: 8⅜ x 3¼" (21.3 x 8.3 cm),
cut vase: 9½ x 3" (24.1 x 7.6 cm)
Gift of the designers, 2004

Sam Hecht
Born 1969
David Sandbach
Born 1963
Chris Chapman
Born 1964

Keyboard, 2000
Pressure-formed ElekTex on siliconized foam
and rubberized silicon with IR transmission
capability, 1¼ x 12 x 5½″ (3.2 x 30.5 x 14 cm)
Manufacturer: Electrotextiles Co. (now Eleksen
Ltd.), Brompton-on-Swale, England
Gift of the manufacturer, 2001

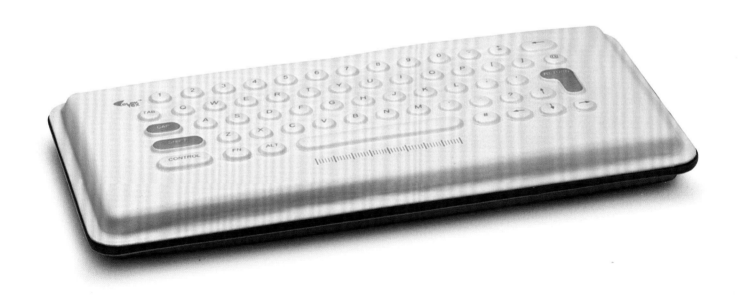

Robert del Naja

Born 1970

Michael Nash Associates

Dates unknown

Cover for the CD *Blue Lines*,
by Massive Attack, 2001
Lithograph and plastic,
4⅞ x 5⅝" (12.4 x 14.3 cm)
Purchase, 2009

Ron Arad

Born Israel 1951

FPE (Fantastic, Plastic, Elastic),
1997
Extruded aluminum profiles
and injection-molded
polypropylene plastic sheet,
31¼ x 22 x 17"
(79.4 x 55.9 x 43.2 cm)
Manufacturer: Kartell, Milan
Gift of the manufacturer, 2000

James Auger
Born 1970
Jimmy Loizeau
Born 1968
Royal College of Art Design Interactions Department
Established 1989

Social Tele-presence, 2001
Acrylic plastic, aluminum, electronic media,
Sony Glasstron glasses, video, and R/C radio receivers,
6½ x 6 x 10¼" (16.5 x 15.2 x 26 cm)
Gift of the Speyer Family Foundation, 2008

Philip Worthington

Born 1977

Shadow Monsters, 2004
Java, Processing, BlobDetection,
SoNIA, and Physics software,
dimensions variable
Gift of the designer, 2008

Amanda Levete

Born 1959

Fruit Bowl, 2005
Laser-sintered nylon,
2½ x 12 x 24½"
(6.4 x 30.5 x 62.2 cm)
Manufacturer: Materialise NV,
Louvain, Belgium
Clarissa Bronfman Purchase
Fund, 2006

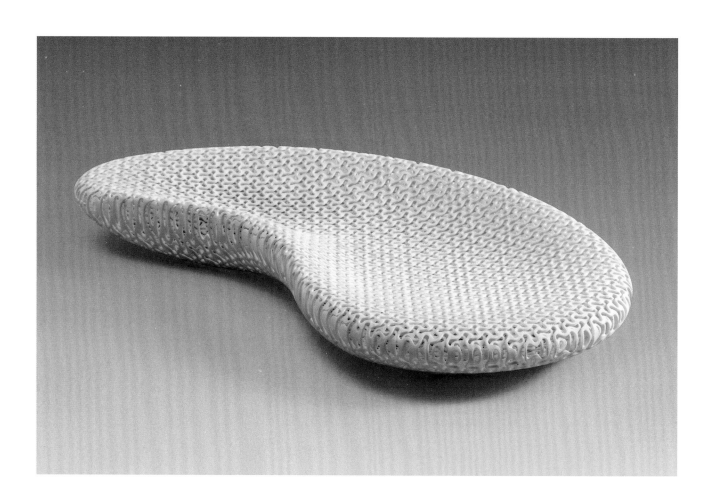

Michael Burton

Born 1977

Nanotopia from the Future Farm
Project (concept illustration),
2006–07
Digital photograph and video
(color, sound), dimensions variable
Gift of the designer, 2008

Lionel Theodore Dean

Born 1962

Tuber 9, 2004
Laser-sintered nylon,
14 x 8¼ x 8¼"
(35.6 x 21 x 21 cm)
Jurg Zumtobel Purchase Fund,
2005

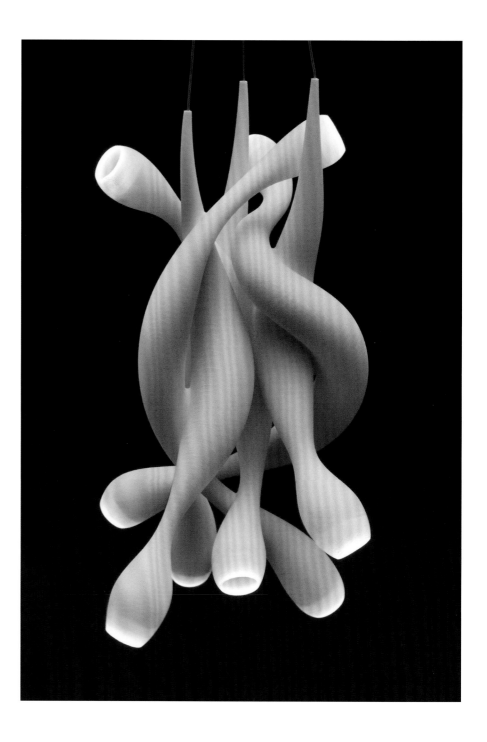

Ron Arad

Born Israel 1951

PizzaKobra, 2007
Chromed steel, aluminum,
and LEDs, extended: 28⅞ x 10¼″
(73.3 x 26 cm),
collapsed: ¾ x 10¼″ diam.
(1.9 x 26 cm)
Manufacturer: iGuzzini
illuminazione SpA, Recanati, Italy
Gift of the manufacturer, 2008

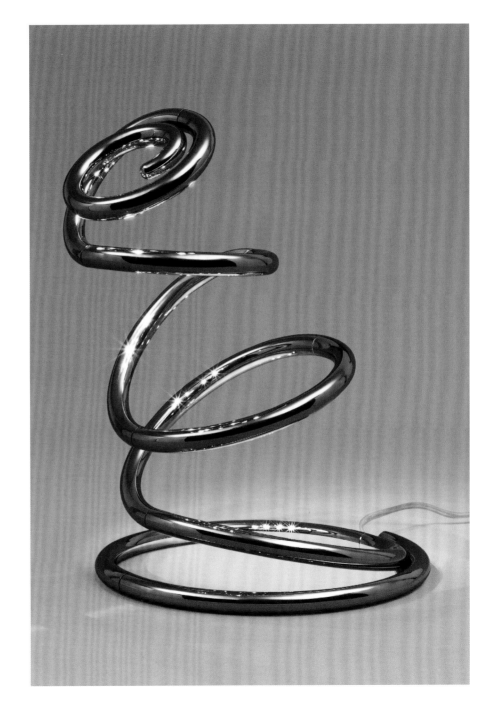

1880–1889	1890–1899	1900–1909	1910–1919

The London shop Liberty, founded in 1875, adds Chinese and Japanese crafts to its catalogue (1880)

Industrialist William Armstrong's Northumberland home is the first private residence with electric light; Gilbert and Sullivan's comic opera *Patience* lampoons the Aesthetic movement (1881)

Industrial design advocate Henry Cole, instrumental in the 1851 Great Exhibition, dies (1882)

The Reform Act extends the vote to sixty percent of adult males (1884)

John Kemp Starley and William Sutton design the Rover bicycle, the first with wheels of equal size and a diamond-shaped steel frame (1884)

William Morris founds Kelmscott Press (1891)

George and Weedon Grossmith's comic novel *Diary of a Nobody* is published, skewering late-Victorian status anxiety (1892)

Queen Victoria's Diamond Jubilee marks the apogee of the British Empire (1897)

Work starts on Charles Rennie Mackintosh's Glasgow School of Art (1898)

China leases Hong Kong and more than two hundred nearby islands to Britain for ninety-nine years; Ebenezer Howard publishes a manifesto for the Garden City movement (1898)

John Ruskin, art critic and champion of Gothic Revival architecture, dies (1900)

Edward Elgar composes his *Pomp and Circumstance* marches (1901–07)

The Second Boer War ends bloodily, contributing to a growing realization among Britons that the Empire is unsustainable (1902)

Britain and France sign the Entente Cordiale, allying to oppose German ambitions in their colonial territories (1904)

Hermann Muthesius publishes *Das englische Haus* (*The English House*), admiring the functionality of British domestic architecture (1904–05)

A Liberal general-election landslide presages social reforms (1906)

Rudyard Kipling wins the Nobel Prize for Literature (1907)

RMS *Titanic* is sunk by an iceberg on its maiden voyage, resulting in the loss of over 1,500 lives (1912)

Suffragette martyr Emily Davison dies after throwing herself in front of King George V's horse at the Derby (1913)

Britain enters World War I (1914)

The British government's Balfour Declaration supports a Jewish homeland in Palestine; painter Norman Wilkinson introduces "dazzle" camouflage for British warships (1917)

World War I ends (1918)

Prime Minister David Lloyd George promises returning soldiers "homes fit for heroes" (1919), but house building does not expand until the late 1920s

1920–1929	1930–1939	1940–1949	1950–1959	1960–1969

1920–1929

The Irish Free State gains independence from Great Britain; the British Broadcasting Corporation (BBC) is founded (1922)

The British Empire Exhibition travels to more than fifty colonial possessions (1924)

A general strike in support of coal workers shuts down industry for ten days; Scottish inventor John Logie Baird demonstrates his mechanical television; Irish engineer Harry Ferguson patents the three-point tractor hitch (1926)

The BBC secures regular public funding (1927)

Sculptor and type designer Eric Gill creates the Gill Sans typeface for Monotype Corporation (1928)

Women win equal voting rights with men (1928)

1930–1939

Aldous Huxley's dystopic novel *Brave New World* is published; Oswald Mosley's Blackshirts gain temporary and sporadic support (1932)

Radio manufacturer Ekco popularizes Bakelite (1934)

King Edward VIII abdicates to marry the divorced American socialite Wallis Simpson (1936)

The Museum of Modern Art, New York, mounts the exhibitions *Modern Architecture in England* and *Posters by E. McKnight Kauffer* (1937)

Designer Reco Capey ends an eleven-year tenure as art director of cosmetics manufacturer Yardley, a role anticipating the influence of American industrial designers (1938)

Britain declares war on Germany, entering World War II (1939)

1940–1949

The British government establishes the Council of Industrial Design, forerunner of the Design Council (1944)

World War II ends; Winston Churchill and the Conservatives lose the general election (1945)

The vast *Britain Can Make It* exhibition presents the nation's consumer goods "to the British people and the world" (1946)

Partition and independence of India and Pakistan begins the dismantling of the British Empire; Swiss typographer Jan Tschichold establishes a graphic language for Penguin Books (1947)

The SS *Empire Windrush* brings the first Caribbean migrant workers to Britain; Kenwood begins manufacture of Chef food mixers; the National Health Service is launched (1948)

1950–1959

The Festival of Britain provides a much-needed injection of national optimism and introduces a characteristically whimsical British modernism in architecture and design (1951)

Queen Elizabeth II ascends the throne (1952)

War rationing comes to an end; meat, butter, cheese, and cooking fat are the last items to be derestricted (1954)

London Transport introduces the double-decker Routemaster bus, designed by Douglas Scott (1956)

The Suez Crisis reveals Britain's weakness abroad; the play *Look Back in Anger*, by "angry young man" John Osborne, opens (1956)

The exhibition *This Is Tomorrow* debuts Pop art in Britain (1956)

1960–1969

Nearly thirty new independent nations are established in the former British Empire

The Cavern Club, Liverpool, hosts some of The Beatles' first performances (1961)

Future prime minister Harold Wilson evokes a new Britain, forged in the "white heat of [scientific and technological] revolution"; the Polyprop chair, designed by Robin Day, goes into production by S. Hille & Co. (1963)

Comprehensive (free and nonselective) schools undergo their greatest expansion (1965)

England beats Germany 4–2 to win the World Cup soccer tournament (1966)

Anarchic comedy series *Monty Python's Flying Circus* debuts on BBC Television (1969)

1970–1979	1980–1989	1990–1999	2000–2009

Irish Republican Army and Loyalist terrorists maintain a state of near civil war in Northern Ireland. An IRA bombing campaign is directed at British cities

Britain joins the European Economic Community, sixteen years after it is founded (1973)

Arab oil-exporting nations impose an embargo, precipitating a widespread economic recession (1973)

The Anglo-French supersonic passenger jet *Concorde* begins commercial flights (1976)

The Sex Pistols release the savagely ironic single "God Save the Queen" (1977)

Margaret Thatcher, a Conservative, becomes Britain's first female prime minister (1979)

Graphic designer Terry Jones, formerly *Vogue* art director, launches the influential style magazine *i-D* (1981)

Britain regains possession of the Falkland Islands after an Argentinean naval assault and invasion (1982)

British Telecom, the state monopoly telephone service, is privatized by the Conservative government (1984)

The annual International Design Conference in Aspen, Colorado, is titled "Insight and Outlook: Views of British Design" (1986)

The Design Museum opens in London (1989)

Aardman Animations' plasticine cartoon figures Wallace and Gromit make their Academy Award–winning film debut in *A Grand Day Out* (1989)

The Soviet Union dissolves, creating new independent states in Eastern Europe (1991)

The Channel Tunnel connects Britain and France; the British National Lottery is established, partially to fund the arts and architecture (1994)

New Labour wins the general election by a landslide; referenda create the Scottish Parliament and National Assembly for Wales (1997)

The Royal Academy mounts *Sensation*, an exhibition featuring the work of "young British artists" (1997)

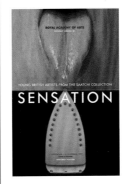

More than a thousand items are selected and exhibited as Millennium Products by the Design Council to represent British creativity (1999)

The London Eye, a giant observation wheel cantilevered over the Thames, is an instant tourist attraction (2000)

New Labour prime minister Tony Blair brings Britain into the Iraq War (2003)

London is selected to host the 2012 Summer Olympic Games; terrorists kill fifty-two people in the 7/7 London bomb attacks (2005)

The British financial sector is heavily damaged by the global financial crisis. The government takes major stakes in some of the country's largest banks (2008)

The exhibition *Ron Arad: No Discipline* opens at MoMA; French designer Philippe Starck fronts *Design for Life*, a BBC reality television show conceived to "shake up the world of British design" (2009)

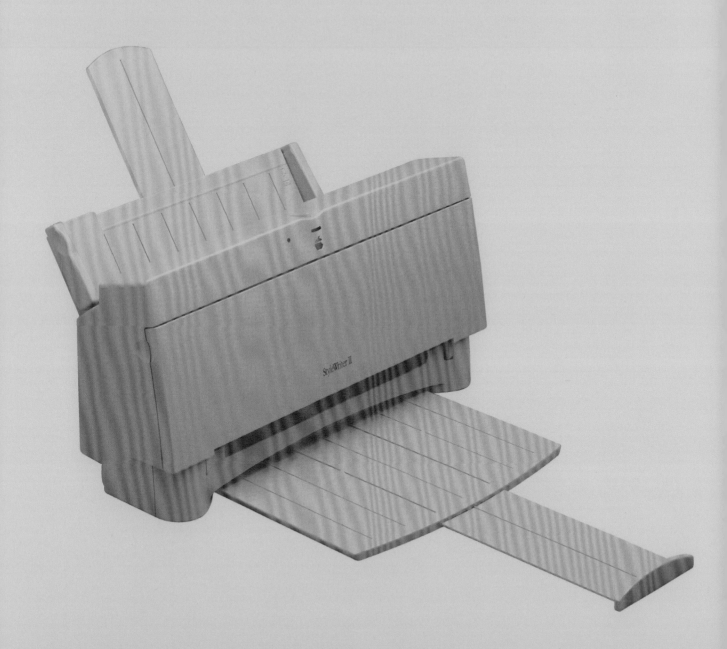

ALDERSEY-WILLIAMS, HUGH. *World Design: Nationalism and Globalism in Design*. New York: Rizzoli, 1992.

BRAGG, MELVYN, and MICHAEL RAEBURN. *Vision: 50 Years of British Creativity*. London: Thames and Hudson, 1999.

David Mellor: Master Metalworker. Sheffield, England: Sheffield Galleries and Museums Trust, 1998.

DUNNE, ANTHONY, and FIONA RABY. *Design Noir: The Secret Life of Electronic Objects*. London: Birkhäuser Basel, 2001.

FORTY, ADRIAN. *Objects of Desire: Design and Society from Wedgwood to IBM*. New York: Pantheon Books, 1986.

FRAYLING, CHRISTOPHER, and CLAIRE CATTERALL, eds. *Design of the Times: One Hundred Years of the Royal College of Art*. London: Richard Dennis, 1996.

GREEN, OLIVER. *Underground Art: London Transport Posters, 1908 to the Present*. London: Studio Vista, 1990.

GREEN, OLIVER, and JEREMY REWSE-DAVIES. *Designed for London: 150 Years of Transport Design*. London: Laurence King, 1995.

HARROD, TANYA. *The Crafts in Britain in the Twentieth Century*. New Haven, Conn.: Yale University Press, 1999.

HUYGEN, FREDERIQUE. *British Design: Image and Identity*. London: Thames and Hudson, 1989.

MACCARTHY, FIONA. *British Design since 1880: A Visual History*. London: Lund Humphries, 1982.

———. *A History of British Design, 1830–1970*. London: Allen and Unwin, 1972.

———. *William Morris: A Life for Our Time*. London: Faber and Faber, 1994.

MACLEOD, ROBERT. *Charles Rennie Mackintosh: Architect and Artist*. London: Collins, 1968.

MARR, ANDREW. *A History of Modern Britain*. London: Macmillan, 2007.

MCDERMOTT, CATHERINE. *Street Style: British Design in the '80s*. London: Design Council, 1987.

MUTHESIUS, HERMANN. *The English House*. London: Frances Lincoln, 2007.

First published as *Das englische Haus*, Berlin: Ernst Wasmuth, 1904–05.

MYERSON, JEREMY. *Gordon Russell: Designer of Furniture, 1892–1992*. London: Design Council, 1992.

———. *IDEO: Masters of Innovation*. London: Laurence King, 2001.

NAHUM, ANDREW. *Alec Issigonis*. London: Design Council, 1988.

OLINS, WALLY. *Corporate Identity*. London: Thames and Hudson, 1989.

PETO, JAMES, ed. *Design: Process, Progress, Practice*. London: Design Museum, 1999.

PEVSNER, NIKOLAUS. *The Englishness of English Art: An Expanded and Annotated Version of the Reith Lectures Broadcast in October and November 1955*. London: Architectural Press, 1956.

———. *Pioneers of the Modern Movement*. New Haven, Conn., and London: Yale University Press, 2005. First published London: Faber and Faber, 1936.

POWERS, ALAN. *Britain: Modern Architectures in History*. London: Reaktion Books, 2007.

POWERS, ALAN, and MORLEY VON STERNBERG. *Modern: The Modern Movement in Britain*. London: Merrell, 2005.

QUENNELL, MARJORIE, and C. H. B. QUENNELL. *A History of Everyday Things in England*. 4 vols. London: B. T. Batsford, 1934.

SPARKE, PENNY. *An Introduction to Design and Culture in the Twentieth Century*. London: Allen and Unwin, 1986.

SUDJIC, DEYAN. *Ron Arad*. London: Laurence King, 1999.

THACKARA, JOHN, ed. *Design after Modernism: Beyond the Object*. London: Thames and Hudson, 1988.

———. *New British Design*. London: Thames and Hudson, 1986.

WALKER, SUSANNAH. *Queensberry Hunt: Creativity and Industry*. London: Fourth Estate, 1992.

WOZENCROFT, JON. *The Graphic Language of Neville Brody 1*. London: Thames and Hudson, 1988.

———. *The Graphic Language of Neville Brody 2*. London: Thames and Hudson, 1994.

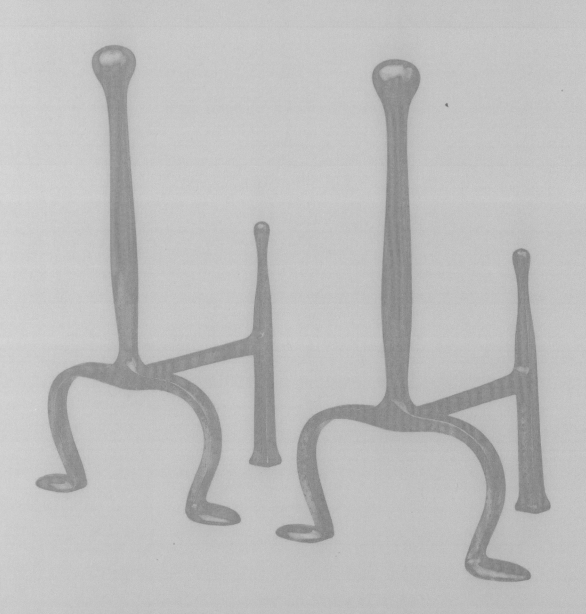

Hugh Aldersey-Williams

Hugh Aldersey-Williams is a writer and curator with interests in science, architecture, and design. He is the author of *New American Design* (1988) and *World Design: Nationalism and Globalism in Design* (1992); part of the latter has been reprinted in *Design Studies: A Reader* (2009). Aldersey-Williams has also written a number of popular science books, including *The Most Beautiful Molecule* (1994) and *Panicology* (2007), and he contributed an essay to the catalogue for the 2008 exhibition *Design and the Elastic Mind*, at The Museum of Modern Art, New York. He has organized exhibitions on contemporary design and architecture at the Victoria and Albert Museum and, most recently, the Wellcome Collection, London, which presented his exhibition *Identity: Eight Rooms, Nine Lives* in 2009. His book *Periodic Tales*, a cultural companion to the chemical elements, will be published by Penguin in 2011. Aldersey-Williams lives in Norfolk, England.

Paola Antonelli

Paola Antonelli is Senior Curator in the Department of Architecture and Design at The Museum of Modern Art, New York, where she has worked since 1984. Since her first exhibition for the Museum, *Mutant Materials in Contemporary Design* (1995), she has organized *Thresholds: Contemporary Design from the Netherlands* (1996), *Achille Castiglioni: Design!* (1997), *Humble Masterpieces* (2004), *Safe: Design Takes on Risk* (2005), and *Design and the Elastic Mind* (2008). Antonelli has taught at the University of California, Los Angeles, and at Harvard University's Graduate School of Design, and has lectured extensively around the world.

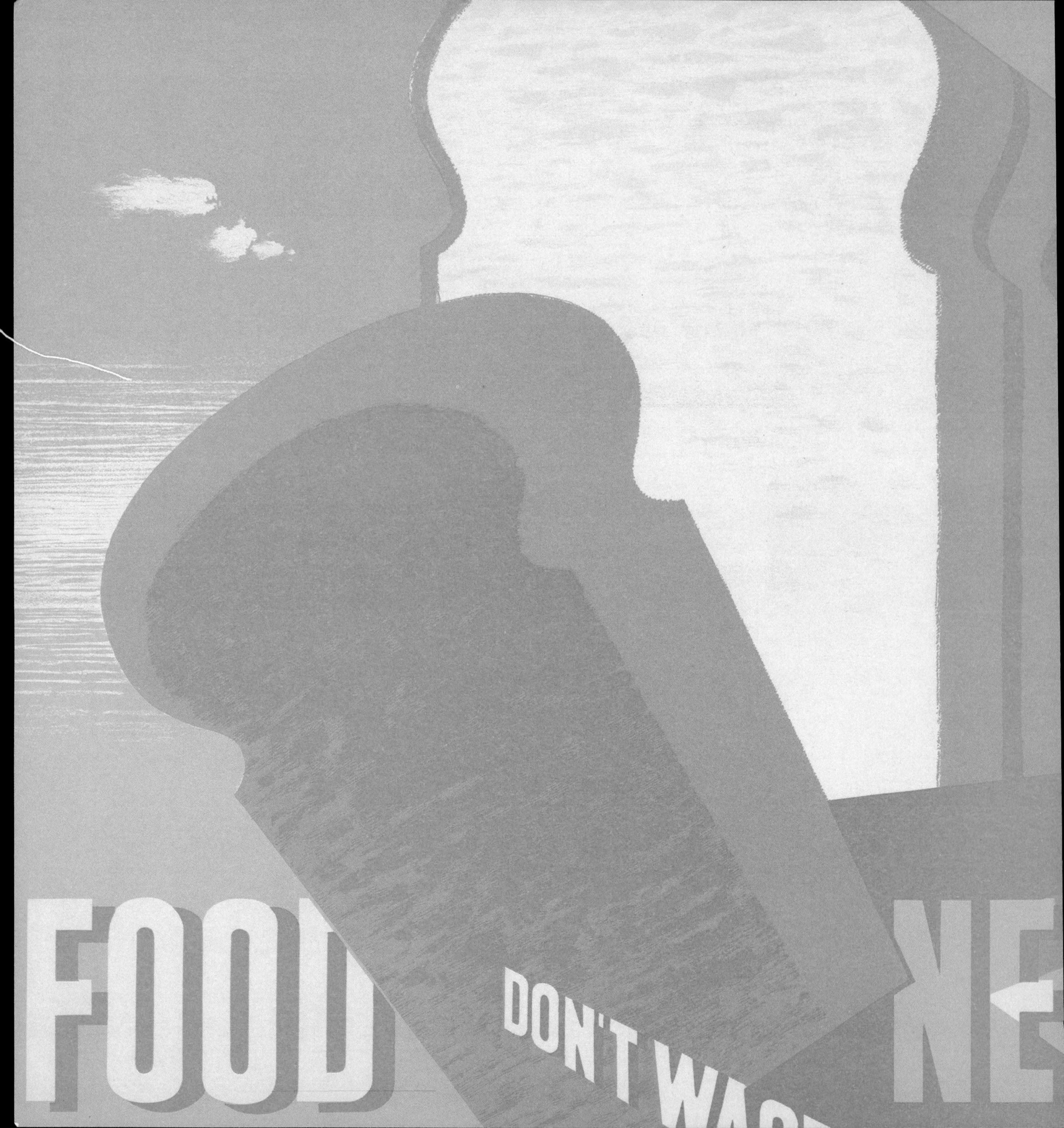
FOOD
DON'T WA
NE